Ocean
Bankruptcy

Ocean Bankruptcy

World Fisheries on the Brink of Disaster

By *Stephen Sloan*

Foreword by *Dr. Sylvia A. Earle*
Afterword by *Richard Reagan*

The Lyons Press
Guilford, CT
An imprint of The Globe Pequot Press

The Lyons Press is an imprint of The Globe Pequot Press.

Printed in the United States of America

Designed by Paul L. Schiff

10 9 8 7 6 5 4 3 2 1

Library of Congress Cataloging-in-Publication Data

Sloan, Stephen.
  Ocean bankruptcy : world fisheries on the brink of disaster / by Stephen Sloan ; foreword by Sylvia A. Earle ; afterword by Richard Reagan.
      p. cm.
  ISBN 1-58574-794-7 (hc : alk. paper)
  1. Fisheries. 2. Fisheries--Economic aspects. 3. Fishery policy. I. Title.
  SH331 .S62 2003
  338.3'727--dc21

                          2002153392

Dedication

This book is dedicated to my grandson Stephen Samuel Sloan II, born on April 21, 2002 at 7:52 PM. It would be my greatest wish for people of his generation that his eyes see the recommendations made here by his grandfather realized during his lifetime.

Stephen Sloan, July 10, 2002

Table of Contents

Exhibits

Foreword

Have you noticed that the price of tuna has risen dramatically in recent years?

Have you wondered why some chefs have chosen to stop serving swordfish?

Have you heard about fights between nations over who has the right to catch cod?

Did you know that prime bluefin tuna can sell in Tokyo for more than $100,000 each?

Are you worried about changes you have noticed in the nature of the sea in your lifetime?

Has it occurred to you that the ocean cannot yield a never-ending supply of wild fish to satisfy an ever-increasing demand?

At times reading like an adventure mystery of danger and discovery, at times like a novel rich with intrigue and laced with heroes and villains, *Ocean Bankruptcy* is told with the conviction and veracity of one who has been and continues to be central to the drama of unprecedented changes now sweeping the world's oceans. In this action-packed saga of trouble and rampant skullduggery on the high seas, of human nature at its best and its worst, of scientific exploration and voracious exploitation, Stephen Sloan tells what few people know or are willing to admit about the nature of fish and, most importantly, the nature of fishermen and the politics that shape and often distort fisheries policies.

Sloan not only knows what he is talking about but also has the courage to speak up and the talent to tell the story of how highly migratory marlin, swordfish, tuna, and other wild fish have gone from natural abundance to cataclysmic decline, and how the disruption and loss of these important natural ocean predators—the ocean equivalent of lions, tigers, and wolves—have rippled through food chains in the sea with alarming speed and disastrous consequences to the ocean, to ocean life, and ultimately, to humankind.

A passionate fisherman, an astute businessman, an eloquent statesman-of-the-sea, Sloan has developed an ethic for the ocean that echoes Theodore Roosevelt's ethic for the land, an attitude that recognizes the need for responsible care for the natural world, to use—but not use up—the living world that sustains us. Sound economic principles provide the underpinnings of Sloan's assessment of the causes of decline of the once abundant sea creatures targeted for taking in recent years. Those same principles provide reason for hope that actions can be taken to help restore and maintain healthy populations of these much admired and greatly endangered creatures.

This book could not be more timely. In the twentieth century, more was learned about the ocean than during all preceding human history; at the same time, more was lost. We learned of the vital importance of the sea to all living things at about the same time that we discovered, to our horror, that the health and stability of the sea are vulnerable to toxic materials we are allowing to flow into it—and to the enormous disruptions brought about because of what we are taking out.

Curiously, many still believe that the ocean has an infinite capacity to absorb wastes and simultaneously yield many millions of tons of swordfish, tuna, and numerous other kinds of fish, no matter how many are extracted. The reality is that a swift, sharp decline of many ocean species in recent years has been coincident with swift, sharp increase in commercial fishing pressures. Pollution is a factor, but there is compelling evidence that nothing has been more damaging to the health of the ocean than the large-scale taking of wild fish and other marine life, compounded by the use of massively destructive gear.

Nets trawled across the sea floor to capture bottom-dwellers not only take targeted creatures, but the entire ecosystem, a technique that can be likened to bulldozing swaths of forest to catch songbirds and squirrels. Sloan describes longlines, sometimes extending fifty or more miles from one end to the other, dangling strings of baited hooks every few feet. The bait alone, whether squid or fish, involves extraction of millions of tons of wildlife from natural systems that have never before known such sweeping losses. The effect is compounded when other creatures—albatrosses and other sea birds, sharks, sea turtles, dolphins, swordfish, tuna, and others—are attracted to the bait and die as "incidental catch." Commercially valuable fish of legal size are kept, but most of what is caught, millions of tons, is discarded—dead. Sloan describes how the numbers of fish caught are significantly understated, and how attempts to alert public consumers to the realities have been suppressed despite alarming consequences to the fish—and ultimately, to the fishermen.

The irony is that people have been fishing to feed their families and local communities for thousands of years, and might have continued to do so had our consumption of wild sea creatures stayed at a reasonable level. But a few decades of modern industrial fishing, powered by sophisticated technologies and driven by new markets and high prices have disrupted millions of years of fine-tuned predator-prey relationships. Nothing in the history of life on Earth has prepared creatures in the sea for the extravagant predation by humankind in recent times.

Less than twenty million tons of wild fish and other sea-life were extracted annually from the world's oceans in the early part of the twentieth century, but following World War II, new demands for seafood, coupled with technologies developed during wartime, resulted in increasingly large takes. Acoustic techniques used to detect submarines could now find every last fish in a population. New lightweight, strong, and inexpensive materials encouraged the development of enormous new nets and increasingly long strings of hook-bearing lines. Precise navigation techniques took much of the guesswork out of returning to favored locations. Soon, the fish had no place to hide. Just as habitat destruction and the taking of thousands of tons of wild birds, buffalo, and other wild animals for commercial markets soon eliminated passenger pigeons, prairie chickens, and Eskimo curlew from North American skies, so did habitat destruction and large-scale taking of fish begin to threaten the very survival of some popular fish species in America's oceans. A similar pattern has occurred elsewhere in the world, first on the land, and now in the sea.

By the late 1980s, nearly one hundred million tons of wild-caught sea creatures were being reported annually. Since then, catches have steadily decreased in number and in size despite massive government subsidies to commercial fishing interests, the deployment of many more fishing boats, increased effort, and the use of new technologies to find, extract, and bring to market numerous species previously regarded as inedible or beyond the range of conventional gear. Policies intended to respond to the problems have often been marred by political pressures from special-interest groups.

All things considered, the politics of commercial fishing—not the fishermen or the markets or the destructive gear—may be the largest problem facing the ocean today. It is this topic in particular that Sloan takes on with special knowledge, skill, and a keen desire not only to set the record straight, but to set in motion new ways of thinking and behaving toward the ocean. Marcel Proust observed, "The real voyage of discovery does not consist of seeking new landscapes, but in having new eyes." Those reading this thoughtful and thought-provoking book will surely be inspired to see with new perspective the nature of fish, fishing, and the enduring relationship between humankind and the sea.

Dr. Sylvia A. Earle
Explorer in Residence
National Geographic Society

Preface

Man's ability to wipe out the fish that inhabit our oceans through his consuming greed and lack of foresight is a never-ending odyssey, so let us begin the journey together. This is complicated by the venality of some politicians and fishery bureaucrats, who for a few coins of the current realm are willing to let these once fruitful waters be made barren. Twelve years ago, I was an outsider. I was a recreational fisherman, considered to be a rich fat cat from New York City, and had no governmental administrative experience. My fishing claim to fame was the setting of forty-four world's records. I served on many boards as a trustee or director, such as the International Game Fish Association, The Billfish Foundation, the National Coalition for Marine Conservation, the American Museum of Fly Fishing, and the Catskill Fly Fishing Museum as well as the South Street Seaport Museum. I also created the *The Fishing Zone* radio show on Talk America Radio Network. I also secured five patents for a floating dock system called the "Infinity Dock System." From this base of activities I worked my way inside to serve our government and its people in fishery matters. My story is about a large part of the Atlantic Ocean governed by the International Commission for the Conservation of Atlantic Tunas (ICCAT).

The oceans of the world, in my opinion, are part of the Public Trust. Supposedly, they are being held by the nations of the world in trust for the world's present and future populations for purposes of commerce, navigation, and fishing. We are led to believe that we are using only the dividends in this account. I will prove that the arguments used by international fishery managers are often fallacious and misleading.

These managers have continuously allowed overfishing and the destruction of both essential fish habitat (EFH) as well as other natural resources such as sea birds, turtles, fish, and mammals. All sorts of ruses are used: Fishing for "scientific purposes" is a favorite to prolong a fishery in spite of assessments that prove that the fishery is in trouble. This tactic has led to one fishery after another being fished in excess of maximum sustainable yield (MSY), the benchmark of the scientists.

The materials and theories offered in this book are controversial, so I offer to those who disagree with me to debate on a full hour of radio on my weekly show, *The Fishing Zone*. Here is my story.

But, before we begin, I wish to thank the many people and instutions that have helped me. The Columbia University Center for Environmental Research and Conservation has enriched this work. The United States government and especially the National Marine Fisheries Service are high on that list because, despite the practices of a few, I have received what amounts to a doctorate education in fisheries. That agency has many faults and is totally biased toward commercial fishing, but I respect hard-working people when they are honest, and there are many such at NMFS and in the commercial fishery industry, but not enough to make a difference. To those from NMFS who play the political game to the hilt at the expense of the world's oceans trust account, I say, perhaps this book will expose those practices and find a better way to deal with international fishery problems. Justice Louis Brandeis said that sunshine is the best disinfectant. To interested parties who care about ocean conservation and the environment, both in and out of the government, this book is meant to be that sunshine.

I wish to thank my editor, James Mooney, who not only did a superb job, but constantly kept encouraging me to complete the work. I wish to thank my wife Nannette Sloan for all her help and encouragement during our forty-five-year marriage. To Robert Sloan, my son, and Suzanne Sloan Doyle, my daughter, thanks for suggesting, "Hey, Pop, you ought to write down your experiences." To their spouses Elizabeth Errico Sloan and Kieran Doyle, many thanks for their unbounded enthusiasm and intellectual curiosity that drive me forward. To Gan Wong, the jigmeister, and Miss Blue, his constant companion, I must thank him for his computer expertise and fishing comradeship.

I have even enjoyed the backbreaking trips, the awful jet lag, the noisy Holiday Inn in Silver Spring where we stay for many of the ICCAT meetings, and eating many of the wrong calories at a wrong time in life because the rewards have been so satisfying.

To the good guys, in order of appearance in my fishery life:

Joe Brooks, John and Kay Rybovich, Ted Naftzger, and Bill Gottwald, all extraordinary fishermen, fishing buddies, partners, and friends. In addition, I wish to thank Guy and Nancy Billups and Sam and Dorothy Evert who stood for everything good in ocean fishing tournaments. A special thanks to Mike Levitt, chairman of International Game Fish Association (IGFA), who broke several of my white marlin light tackle records and kept both of us smiling. To Dr. Ben Sherman and his son Gary who taught me the vagaries of the Hudson Bite. To Richard Reagan, who funded many a fledgling fishery project through his outstanding leadership at the Norcross Wildlife Foundation, and to his wife Sonya Manzo Reagan (Maria on *Sesame Street*) who let her daughter Gabriella go on wonderful fishing trips with us at a very young age.

To Bill Fox who gave me a chance to be a civilian leader in fishery affairs at NMFS. To my dear friends Barbara and Rollie Schmitten whose respect is one of my most prized possessions. Rollie was former head of NMFS, ICCAT commissioner, United States Whaling commissioner, and Atlantic Salmon commissioner and is now director of habitat at NMFS. His insight, his untiring efforts to bring compliance to ICCAT's contracting parties is for too long an unsung anthem. To Kimberly Blakenbeker, secretary of the American delegation to ICCAT, my thanks for being a tireless worker on behalf of the United States fishery interests. A cheery smile always adorns her countenance.

To Dave Balton, from the United States State Department, who never went to an ICCAT meeting unprepared and was indicative of the very best of America. To Will Martin, former United States commissioner, my thanks for your brilliant negotiating and unruffled demeanor when the going got rough. My best wishes to Bill Hogarth, new head of NMFS, straight-shooter, and now head commissioner for the United States at ICCAT. To all those guests I have had on *The Fishing Zone* radio show. Thank you for getting up at ungodly hours every Saturday morning and providing both the public and myself greater insights into fishery matters. There are many other "good guys" who are serious about preserving the oceans including Mary Barley,

Jimmy Donofrio, Tom Fote, John Koegler, Phil Kozak, Skip Walton, professor Bob Ditton, Jim Chambers, Bob Eakes from North Carolina, Ken Hinman and Chris Weld from the National Coalition for Marine Conservation, and Captain Joseph McBride, my comrade in arms against a sea of fishery troubles and a true friend. To my muse, my dear friend, and the greatest marine painter who ever picked up a brush, Stanley Meltzoff, who taught me more about art, painting, composition, and artistic endeavors than I could have ever learned in any university; I thank him for all the hours he left his easel and spent with me. John and JoAnne Payson were constant supporters of conservation causes and good friends of the world oceans.

To those who have had other agendas, both public and secret, you know who you are and some are portrayed in this book.

Because this book is a work in progress and the user nations continuously work at reducing our trust fund of fishery stocks, I will regularly update the book on my web page after its publication. You may also obtain a CD or videotape of a graphic film showing longlining at its worst.

Send $5.00 for each copy to The Fisheries Defense Fund, Inc., 1040 First Avenue, Suite 367, New York, NY 10022. We have enclosed a tear-out card for your convenience.

This book should also be regarded as an early warning to the subsequent events of September 11, 2001, regarding fishing vessels under flags of convenience as potential vehicles of terrorism. And so, I begin by taking you on a trip demonstrating the inside fishery game as our oceans are being stripped of their very life and our wealth.

List of Acronyms and Abbreviations Used in Text

ASA	American Sportfishing Association
BAYS	Bigeye, Albacore, Yellowfin, Skipjack Tunas
BET	Bigeye Tuna
BFT	Bluefin Tuna
BWFA	Blue Water Fishermen's Association
BYT	Bigeye Tuna
CAACC	Confederation of the Associations of Atlantic Charter Boats and Captains
CCA	Coastal Conservation Association
CCP	Contracting Parties
CEO	Chief Executive Officer
CITIES	Committee for the International Trade in Endangered Species
COOP	Cooperating Parties
EC	European Community, called European Union following Maastricht Treaty of November 1993
ECTA	East Coast Tuna Association
EEZ	Exclusive Economic Zone
EFH	Essential Fish Habitat
FAD	Fish Attracting Device
FISI	First In Stays In
FOC	Flag of Convenience
FPV	Factory Processing Vessel
FRB	Fish Recovery Bond
F/V	Fishing Vessel

GPS	Global Positioning System
HMS	Highly Migratory Species
ICCAT	International Commission for the Conservation of Atlantic Tunas
IGFA	International Game Fish Association
IPO	Initial Public Offering
IUU	Illegal Unregulated Unreported
JCAA	Jersey Coast Anglers Association
MAFAC	Marine Fishery Advisory Committee
MBA	Masters of Business Administration
MBCA	Montauk Boatmen and Captains Association
MITI	Ministry of International Trade and Industry
MOU	Memorandum of Understanding
MSY	Maximum Sustainable Yield
NCMC	National Coalition for Marine Conservation
NFA	National Fishing Association
NFI	National Fishery Institute
NGO	Nongovernmental Organization
NMFS	National Marine Fisheries Service
NOAA	National Oceanic and Atmospheric Administration
OD	Overdose
PWG	Permanent Working Group
RFA	Recreational Fishing Alliance
SCRS	Standing Committee on Research and Statistics
SWO	Swordfish
TBF	The Billfish Foundation
TRO	Temporary Restraining Order
VMS	Vessel Monitoring System
YFT	Yellowfin Tuna

CHAPTER 1

United States Federal Court, Second District
Hauppauge, New York
October 12, 1991
1000 hours

• •

Miguel Alcoron lives in Faial, the Azores. He is a sixth-generation Portuguese fisherman who has plied the once rich waters surrounding such islands as Faial, Pico, São Jorge, Flores, and the distant capital island, São Miguel.

Miguel works on the skipjack tuna bait boat *Lucinda* docked at the commercial wharf in the harbor of Faial. She is forty years old and twenty meters long, and is powered by a single 400-horsepower Mans engine. She cruises at eight knots in all kinds of weather except gale-force seas. On her deck she has a rubber hose, two and a half centimeters in diameter, circumnavigating the deck and fastened to the top of her covering board. The hose is punctured every twelve centimeters by a

hole and is attached to a pump. Stacked up against the wheelhouse are twenty five-meter-long bamboo poles, each with a
distinctive mark made by Miguel and the other fishermen. On
top of the wheelhouse are two ten-meter-long bamboo outriggers that are mounted so that they can be lowered to provide a
forty-five-degree angle to the vessel. The outriggers have a pul-
ley system and a clip to hold the Dacron trolling line that is set
astern ten boat-lengths (two hundred meters) from the vessel.
Attached to the trolling lines that are strung through the outriggers is a mother-of-pearl jig, just about the same size and shape,
ten centimeters, as the local bait, a small mackerel fish called a
chicharo that is kept alive swimming in an aerated tank. The
Azorean skipjack tuna fishermen have used this system for over
a century.

At daybreak Miguel begins walking from his home on the
slope of Faial's hillside down to the port of Horta. He keeps
checking the sky for weather and the signs are good. A light
southwest breeze will make the ocean flat; there are no low-
flying cumulus storm clouds racing in from the northwest on
this sweet and cool early July morning. The reports gleaned
from other fishermen say that a high weather system will hover
around the Azores for a week. Miguel has a good feeling about
this coming voyage. He looks at the island of Pico ten kilometers across the bay. He cannot miss the now silent, now snow-
covered volcano peak, rising 2,350 meters above the seas. It is
Pico's crown and fisherman's friend, for it can be seen sixty
kilometers out to sea.

Miguel reaches the protected harbor and heads for the
Lucinda where he takes a cup of black, rich, hot coffee in a bat-

tered but still serviceable mug. The engine coughs and the *Lucinda* heads out of the harbor and then steams west-north-west for almost three hundred kilometers for the island of flowers, Flores. There is no anchorage at Flores, but the slopes of the hills are as if Van Gogh and Gauguin were turned loose to run riot with color over the landscape.

The fishing begins. Miguel is a catcher. The mother-of-pearl jigs go over the side. Flores is now some twenty kilometers to the northeast. The *Lucinda* slows to a trolling speed of four knots and takes a gentle tack. After only twenty minutes both lines snap out of the outriggers with a crack and become taut. The boat has hooked two skipjack tunas weighing three kilos each (they are *Katsuwonus pelamis* of the *Scombridae* family of fishes). These fish are among the smallest members of the tuna family, rarely weighing more than twenty-two kilos. Skipjack tunas are a schooling, pelagic, migratory deep-water species. They feed near the surface, forming schools of over fifty thousand individuals.

Two seamen grasp the lines as the boat slows down to two knots, and they begin hauling the fish in hand over fist. When the fish are astern, down about three fathoms, the two lines are secured to the port and starboard hawsers. The skipjacks flash like refracted mirrors in the cobalt blue water. The water hose is turned on and a spray leaps overboard, hitting the ocean's surface. At the very same time another crew member begins brailing, or ladling, live *chicharos* overboard. The appearance of a feeding frenzy is thus created, attracting other skipjack tunas that usually travel in massive schools. It works; within fifteen minutes the water is boiling with a feeding frenzy.

Miguel takes his pole and attaches a barbless hook and a feather to the five meters of line running from the tip. Sometimes he elects to use live bait, but today there are many, many skipjack tunas alongside the *Lucinda*. He casts the feather into the frenzied mass and the line becomes taut immediately. The fish surges downward, but in less than ten seconds the line is taut against the pole and the skipjack, weighing about two kilos, begins to spiral up to the surface, losing the battle against Miguel's unforgiving arched pole. The moment that Miguel can break the fish's head out of the water, on the next beat of the tail, and with the expertise of all his preceding generations, Miguel flips the pole skyward and the tuna becomes airborne and lands with a thwack on the deck behind him. With an imperceptible and lightning-quick half-turn of his wrist, he dislodges the barbless hook and flips his lure forward again to repeat the process many, many times before sunset. Crew members shovel the skipjacks now on deck, their tails beating sweet rhythms into Miguel, into the unrefrigerated hold of the vessel, working furiously to keep up with the pace of Miguel and his fellow fishermen.

Miguel notices that on the second day, he seems closer to the water line, and on the third there is little freeboard left for the heavily laden vessel. It is time to go home and unload the catch. Within fifteen hours the snowcapped volcano at Pico is in sight, beaconing the ship and its crew home to Faial. The *Lucinda* turns into the calm water behind the breakwater and heads for the commercial dock. A squadron of gulls follows the boat, diving for scraps washed overboard by the crew scrubbing the deck for the next-to-last time on this voyage.

The *Lucinda* makes fast to the commercial dock. All hands and a crane crew help with the unloading. The skipjacks are packed in fifty-kilo cartons, and a thin layer of ice is thrown on top. They are then taken to the icehouse and will be shipped by airfreight to a cannery at Lisbon or the auction house in Madrid. This is the first refrigeration the catch has seen in three days of fishing. The *Lucinda* begins to rise at the water line as the catch is packed.

Miguel helps in the unloading, knowing that he will soon have a share, a pay day, and be able to spend a few days with his wife and four children. A crane suddenly places a large steel container next to the vessel. The skipjacks left in the hold, approximately one-third of the catch, suddenly take on a stranger new property. They have become gurry, or gruel, from being crushed by their successors. These final fish were Miguel's first efforts, and are now being unloaded like thick soup into the metal containers. Plastic liners are placed in the containers. This last byproduct of the fishing trip will be given away or sold for pennies to the local farmers who spread it on the sloping fertile fields of Faial as fertilizer for the crops and flowers that proliferate there. Miguel has no thoughts on this final process. He picks up a hose and begins to wash down the boat, getting ready for the next skipjack tuna voyage.

Far from Miguel's homeport, massive catches of small skipjacks of about two kilos have been reported in the Gulf of Guinea off West Africa. Miguel is aware that there are larger boats than his and has been told, through the jungle drums of a fisherman's waterfront, that the French and Spanish have built many purse seine vessels sixty meters long holding a hundred times the capacity of the *Lucinda*. These boats are completely refrigerated. He even heard one incredible rumor (sadly true) that the shipyard in Vigo, Spain, built two 120-meter purse seine vessels, with a capability of 3,000 metric tons in the hold. He has even heard of an organization called ICCAT, although he does not know its structure or how it works or if anything done so far away will affect him and his family.

Miguel is smiling. He is finished, he's been paid, and he is heading home. As long as the *Lucinda* sails, he is happy. Little does he know that his fate as a fisherman is not in his hands. It is in the hands of the International Commission for the Conservation of Atlantic Tunas (ICCAT). Details of its organization and structure are found at Exhibit A at page185.

In 1991, Carmen Blondin was governmental head of the American delegation at ICCAT. He was assisted by two commissioners from the private sector: Lee Weddig, head of National Fishery Institute (NFI), a commercial lobbying group representing processors, purse seiners, tuna dealers, and fishery importers and exporters; and Mike Montgomery, a lawyer and fundraiser for Presidents Reagan and Bush, who hailed from California, as representative of Eastern recreational fishing interests.

All of us in the East had our suspicions about why Reagan appointed someone from the shores of the Pacific in California to become the spokesman for Atlantic tuna and swordfish interests. At its worst, Montgomery's appointment was just another slap in the face to the Atlantic and Caribbean recreational fishing interests. At its political best, it was a poor payback—low-level appointment for a fundraiser to visit Washington and go to the "Hill" to lobby for his clients on the taxpayers' money. This

American ICCAT meeting was held in early November in Washington, in order to "form" United States policy based upon comments and suggestions from all user advisory groups during the public comment period, as required by law. All this was done just before the main international meeting in Madrid. In between was the trilateral meeting held in Ottawa, Tokyo, or Washington, where the United States, Canada, and Japan meet with only governmental representatives in attendance. The commercial representative Lee Weddig and his successor Glen Delaney often attended at the invitation of the American commissioner.

To my knowledge, no such invitation has ever been extended to a recreational commissioner nor has one ever attended. It is at the trilateral that discussions about future strategy for the upcoming international ICCAT meeting are held. It is here where the "deals" are usually cut, or at least the disagreements are on the table so there is no surprise for these three delegations in front of the other countries in the open plenary sessions. As a prelude to the 1991 trilateral Ottawa meeting, Blondin had just come back from Hauppauge, Long Island, where he testified in a court case that I had started on behalf of the Montauk Boatmen and Captains Association (MBCA) whose tuna season at the beginning of October 1991 was summarily closed by the National Marine Fisheries Service (NMFS), but left open to commercial fishermen (purse seiners, harpooners, and pin-hookers and longliners). A pin-hooker is a commercial fisherman using a conventional rod and reel as gear. NMFS claimed that the recreational interests had reached their quota. MBCA claimed that NMFS did not have the correct numbers. I had gotten a call from Captain Joe McBride, president of MBCA, who begged me to get

7

NMFS to reopen the season. He called me because I was on the American delegation and had vast experience in world record saltwater fishing. I explained to him that it would be difficult, almost impossible, to get NMFS to reopen the season, but a better chance might be to get a Federal judge to close the season for everyone because the NMFS data was not only flawed, but incomprehensible. The plaintiff was the MBCA and the defendants were Secretary of Commerce Robert Mosbacher, Assistant Secretary of NMFS Bill Fox, and the National Marine Fishery Service. The intent of the suit was to close the season for everyone until NMFS could sort out the correct catch data pertaining to the quotas issued by NMFS in conjunction with the process at ICCAT. I told Joe McBride I would help to bankroll the lawsuit.

All this happened before the ICCAT commissioners Carmen Blondin, Lee Weddig, and Mike Montgomery went to Madrid.

We had asked the judge to close the season for everyone by issuing a temporary restraining order (TRO) until NMFS could provide us with the data showing just how much bluefin tuna had been caught by all of the user groups. During the court proceedings, the judge looked down from the bench at our attorneys and me, and asked, "What is the United States quota?" to which Blondin replied, "1,450 metric tons." Then the judge asked, "How much has been caught, according to NMFS?" Blondin said, "950 metric tons." The judge, looking at us over his glasses, said, "How can I close a season when there is still quota to catch?" The MBCA lawyer, Scott Furman, said, "Judge, they haven't the foggiest idea of how much has been caught." The judge ruled, "I am not going to close the season; however, the government has to produce all of its numbers in ten days. So ruled." Bang went the gavel.

Later, in a diner over supper and a beer, all from the MBCA were disheartened by the turn of events. "You can't fight City Hall," was one remark, to which I replied that "all was not lost." I reminded those present that this is the first time in the last fifty years that NMFS has had to produce any numbers support-ing its actions. I'll bet the staff would have better luck throwing a dart at the telephone book. This was on a Tuesday night. On Friday morning, a Department of Justice lawyer called our lawyers, Scott Furman and Peter Shatzkin, to advise them that a funny thing happened on the way to obtaining the numbers under the judge's ruling. Since Wednesday, a period of only two days, some 450 metric tons had been discovered to have been caught. There was a miscalculation of discards, the Justice Department lawyers said. Under the Atlantic Tunas Conservation Act, the basis for the United States's joining ICCAT, discards had to be counted. Discarding is the process of throwing back undersized and/or over-quota fish.

We got the judge out of bed on Saturday morning for a hear-ing on issuing the temporary restraining order. The judge asked, "What has happened for me to devote a Saturday to this?" The lawyer from Justice replied, "Your Honor, another 450 metric tons [nearly one million pounds] have been discovered and caught." With astonishment the judge said, "You mean to tell me that 450 tons have been caught in just two days?" To which the lawyer from Justice replied sheepishly, "Your Honor, under the treaty an allowance for dead discards has to be taken." The judge then asked, "Isn't a dead fish a dead fish?" but the lawyer from Justice gave no answer.

The judge spoke, "I am issuing an order for all vessels fishing for bluefin tuna to return to port by midnight Sunday. I am scheduling a hearing in my courtroom for next Friday, ten o'clock in the morning, six days hence. My advice for you [pointing to the Justice Department lawyer] is to read the treaty. I want the cooperation of the Coast Guard and NMFS here. Do I make myself clear?" And so it happened; Atlantic bluefin tuna fishing was brought to a standstill amid the curses of the commercial fleet from Gloucester, Massachusetts, to Cape May, New Jersey, and beyond. Most of the curses were sent in the direction of Captain Joe McBride and me. The first time the NMFS staff had to face a court order to produce their records, they folded. On Sunday at midnight there were no vessels fishing for bluefin tunas on any grounds from Maine to North Carolina. Our strategy had worked, or so we thought.

As soon as I pulled into the parking lot six days later with Joe McBride and Scott Furman, I sensed that something was different. There were twenty large buses, with Massachusetts license plates, all gleaming silver in the sunlight. Disgorging from the buses was a band of rough-looking hombres. None of them had a waist size under 48. All of this mass of humanity headed toward one spot, the judge's courtroom. Inside, there was an elaborate security system and, to make us feel more at ease, forty-four federal marshals, all with guns, patrolled the halls of justice. To add to the melee, twenty electricians were running around trying to rig up all of the extra courtrooms for the obvious overflow. There was mayhem in the air and murder on the minds of some of the rotund visitors from Massachusetts. The number of men far exceeded the legal number of purse seine and harpoon permits issued to fish for the bluefin tuna. We began unpacking our legal papers that made a pile whose size had risen to four feet high. This was amid some comforting remarks like: "We eat guys like you for breakfast," or "Come on my ship, bub, it'll be the last fishing trip youse'll ever take," or "We know how to harpoon guys like you." But the top remark came from Jerry Abrams, head of the syndicate and East Coast Tuna Association (ECTA) that controlled the purse seine boats. Jerry looked me in the eye and said with a mist in his own, "I am doing this for those wonderful guys out there. It's not about the money." We all know what Senator Trent Lott said, when someone said that it's not about the money; it is totally about the money. I told Jerry that he was full of shit. He turned his back and walked away.

The judge came into the courtroom. The next fifteen minutes went by like a trance. I must give the Department of Justice credit here; it got every government witness to give false testimony but got away with it.

The judge asked, "What does the treaty say about discards?" to which Blondin responded, "Judge, it allows discards not to be counted [big lie number one]." Then the judge said, "If discards do not count [apparently the judge forgot that he himself said, 'Isn't a dead fish a dead fish?' and he continued], "I am revoking the TRO. Fishing may continue until another 450 metric tons are caught." Down came the gavel. It dawned on me what had actually happened. The judge had taken testimony from government officials and never once did they ever raise their right hands and place their left hands on a Bible. Therefore, there was no perjury in the case. Blondin had lied about the discards.

The case was over. As we left the courtroom, I heard Blondin say, "I'll get those bastards, I will," looking straight at us. It was a peculiar remark because he had won the case, based upon his incorrect testimony about discards not counting under the treaty and under ICCAT. They do count. So much for the oxymorons, honest courts, and honest government.

A quirk of nature happened right after the trial. Two storms blew in over the Atlantic. The second was the infamous Halloween Storm or "The Perfect Storm." The tunas scooted for warmer climes and the end of the bluefin tuna fishing season was a bust. I figured we saved 400 metric tons of tuna that year because of the court case. Little did I realize how costly a victory that would be.

CHAPTER 2

Japanese Delegation Room
Hotel Pintor, Madrid
November 1991
1200 hours

• •

Kuzio Shima, head of the Japanese delegation and senior ranking member of the department of fisheries in Japan, and Blondin of the American delegation, arranged a lunch to discuss the problem of the catches of small tunas by American recreational fishermen.

Mr. Shima is a wiry man; he holds his body to his left side and walks with a pronounced limp. Do not be mistaken, though, for he is a man of steel with a keen intellect. He is fiercely xenophobic. He believed, like Ishihara's book, *The Japan That Can Say No*, that Japan can say no and must say no to the United States. He was also the Japanese whaling commissioner. It was on his watch that Japan actually expanded its commercial whaling activities under the guise of science. I met him personally in November of 1994. My son, who lived in Japan for seven years and also worked for the Ministry of International Trade and Industry (MITI), set up the meeting. I had heard that the Japanese had been at work raising bluefin tunas in captivity, not for food, but for a biomass. I had heard there was a bluefin experimental fishery project somewhere on a tiny island near Taiwan.

I walked into Shima's office at the Ministry of Food, Agriculture, and Fisheries. It was typical of a bureaucratic office, plastered with posters, in this case showing life cycles of fish. Shima had several of an abalone in his modest office. I began to realize that this office was quite spacious according to Japanese standards where three people share a desk in the bullpens of the Japanese ministries. Shima asked in good, but not perfect, English, "What is your interest in bluefin tunas?" "I have a dream," I replied. "I believe we could restock the oceans with highly migratory species like bluefin tunas and bring the overfished stocks back to health." "That is a very worthwhile dream," he answered. "If you would be prepared to visit Japan next summer, I will arrange a trip for you to visit the island of Ishigaki. It is the last holding Japan has below Okinawa." "Absolutely," was my one-word answer. Then he said something quite curious. "You know there will always be some hole in the vast oceans where a group of bluefin tuna will spawn." But I am getting a little ahead of my story.

Back to the American commissioner, Carmen Blondin, in Madrid. Shima and Blondin, together with the Canadian representative, arranged a lunch to discuss the problem of small, school-size bluefin tuna catches by United States recreational fishermen. Actually, before Blondin raised the "problem," there was none, but he created one. U.S. recreational fishermen were limited to 15 percent of the total quota for the Western Atlantic. The quota was 2,600 metric tons; therefore they could catch 390 metric tons, more than three-quarters of a million pounds, but not sell them.

The tunas were only for personal use, home consumption. This catch was the backbone of the charter boats, the "six-pack" fleet so named because the Coast Guard license restricted them to six persons plus two in crew, and recreational private sportfishing boats. The charter boats would arrange tuna charters by manning the booths at the various sporting and boat shows between December and April of any given year. Bookings and deposits were arranged months in advance. These charters brought the highest prices in the business and deservedly so, for if you ever sat through one of the many winter outdoorsman shows day after day booking your charters, you have earned every penny the business brings. No one complained. The quota seems fair, and the recreational fishermen never fished over that amount and for many consecutive years were under the quota. Charter boats did not target the big money giant fish. A bluefin tuna weighing over five hundred pounds and about six to eight years old, is ready for spawning. Because it is also sought after as the prime ingredient for sushi, these giant bluefins are worth at the dock from $7.00 a pound to over $20.00 a pound, depending on the marbleized fat content of the meat. Do the math. A fish of seven hundred pounds could be worth $4,900 to $14,000. There is on record one fish that sold for $80,000; however, that was recently topped in January 2001, when a 400-pound bluefin brought $172,000 or $430 a pound in Japan's market. This fish, I am told, was sold in a very scarce market with high demand.

Mike Montgomery, the 1991 U.S. recreational commissioner, arrived at the luncheon meeting in Madrid and was told by Blondin that the Japanese had not ordered a sandwich for him, so he was excused from the lunch. This is curious, because Blondin's secretary at the time, Ms. Becky, found a way to wangle a cheese sandwich from the Japanese delegation and she went into the meeting. Montgomery was furious and went to his hotel, packed, and checked out for California. Blondin went in and negotiated a new deal for the U.S. recreational fishermen without their appointed representative in attendance.

In 1992, U.S. recreational fishermen would have had 15 percent of the combined U.S.-Canadian-Japanese quota for the Western Atlantic, which was 2,600 metric tons. This amounted to 390 metric tons. The entire Western Atlantic quota was shared as follows:

United States	50% =	1,300 metric tons
Canada	35% =	910 metric tons
Japan	15% =	390 metric tons

No one has ever questioned why Japan is entitled to 15 percent of a fishery quota when the fishing grounds are halfway around the world from Japan's fishery economic zone. No such reciprocity exists in the home waters off Japan, where fishery after fishery has collapsed. I imagine the reasoning follows the business adage, "One does not aggravate one's best and only customer."

Blondin made an agreement with Japan and Canada to give the U.S. recreational fishermen 8 percent only of the U.S. quota, or 104 metric tons, or about a quarter-million pounds. This was all done without our recreational representative present and, in effect, sent home. This amounted to nearly a 75 percent cut in quota for the American recreational tuna fishermen. Montgomery was not there to protest. Canada did not catch, market, and sell the small fish and therefore felt that if about 300 metric tons of tuna could grow up one day, they would become valuable to Canadian commercial fishermen. Out of spite, meanness, and pique, Blondin got more than even. He cost small-business owners, the charter fleets, the mom-and-pop motels, the tackle shops, the gas stations, and the marinas along the Atlantic coastline from Maine to Texas tens of millions of dollars. No one, especially Blondin, ever checked the Small Business Administration Act to see if their new deal was legal.

CHAPTER 3

The White House
Washington
December 1991
1200 hours
(Simulated conversation based upon
conversation with others)

• •

G.H.W. Bush:	Get Knauss [Dr. John Atkinson Knauss was head of the National Oceanic and Atmospheric Administration (NOAA) and in charge of NMFS and Blondin] on the phone right away.
Staff:	Yes, Mr. President.
Bush:	Knauss, I just got a call from George Hommel, George Barley, and Curt Gowdy about Blondin. What the hell happened over in Spain? Did he really give away three-quarters of the American recreational bluefin tuna catch? Jesus, there

are a hell of a lot of recreational fishermen who voted for me in the last election. What does Fox say about this? [Bill Fox, head of NMFS, was appointed by President George H.W. Bush at the suggestion of George Barley, George Hommel, Curt Gowdy, and Perry Bass, the so-called et-com committee or, in other terms, his fishing buddies and special advisers on fishery matters.]

Knauss: Blondin is mum about an explanation. He says the science is on his side.

Bush: Bullshit. I tell you what. I want Blondin on the road explaining to as many recreational fishing groups as will listen to his version of what happened. You know I cannot, under the Magnuson-Stevens Act, revoke what ICCAT ruled on over there in Madrid. We're stuck with Blondin's decision. Knauss, keep him on the road until this dies down. And, Knauss, next year before he goes to ICCAT, I want a meeting with him to tell him what the American agenda is before we do something stupid like this again. Is that understood?

Knauss: Yes, Mr. President.

So, Blondin went on the road. He tried to blame what happened on Japanese bad manners in not ordering a cheese sandwich for Mike Montgomery, but it was a pathetic road show. He kept reminding everyone that he was the president's aide at the

United Nations when the president was our ambassador. He more-than-slightly hinted that his tenure as Bush's aide was laced with overtones from the CIA. He called out his war record (the Coast Guard) and his family history of war service. The veterans in the audience were not impressed and it made him seem a much smaller person than he already was. The road show ended mercifully after two months. The 8 percent quota is still with us in the year 2002, eleven years later.

Notes: George Hommel is a guide from the Florida Keys. He also owned Worldwide Sportsman Tackle Shop and Guide Service in Islamorada. He recently sold it to Johnny Morris's Bass Pro Shops. He is a fine gentleman, a good friend of mine, and has taken the Bush family out fishing for years. He also fished with Curt Gowdy of Red Sox and radio and television fame. Curt was also the host of *American Sportsman* on ABC-TV, the finest fishing TV show ever made. Curt is a fellow IGFA trustee.

George Barley died in a plane crash on the way to meet with the Army Corps of Engineers to discuss the Everglades and how to save this critical area from destruction. George had proposed a one-cent-per-pound rebate from the sugar growers to help pay for the restoration. The FBI is still investigating the crash. Alfie Fanjuil, head of the family that controls much of the Florida sugar production and a recipient of a ten-cent-per-pound subsidy from the United States government, opposed Barley's plan, as well as any contribution by industry to help. Alfie felt it was a federal problem, not a sugar industry one. He may be noted for his foresight because the government finally, under President Clinton, approved seven billion dollars in the year 2000 to restore the Everglades. This was a hundred times the amount of rebate contemplated by George Barley. George's widow Mary is doing great work following up the Everglades restoration and battling the Corps of Engineers. Assisting her is Paul Tudor Jones and Everett Ehrlick.

The annual rainfall in Florida is sixty-four inches. The Corps of Engineers releases sixty-two inches out into the ocean to keep the sugar land dry to allow the sugar growers to make more and more money. Ten cents per pound is a subsidy by the United States government to combat Castro's sugar prices.

Perry Bass needs no introduction. He and his family are powerful investors in Texas, and Perry, very quietly, was the banker, along with Joe Cullman from Philip Morris, who put up the money to begin the lawsuit to stop the netting of Atlantic salmon on the high seas. It is a largely unsung story of huge proportions and benefit to this wonderful fish. If the Atlantic salmon is caught in Canadian rivers by anglers halfway through the twenty-first century, it is primarily due to Perry Bass, Joe Cullman, Lee Wulff, Orri Vigfussen of Iceland, and Rollie Schmitten from our government.

CHAPTER 4

Block Island Sound
Southwest Corner of the Dumping Grounds
Loran Numbers 14400/43600
August 11, 1993
Fishing Vessel Purse Seiner Connie Jean
0900 hours

. . . .

On August 11, 1993, Captain Al Anderson knew he was going to have a bluebird tuna day. He sniffed a gentle southeast breeze and, although it was still dark, the Weather Channel reported a slightly overcast sky. Perfect conditions, he said to himself. The six anglers in his charter stepped aboard. The engine coughed and they were on their way. The stars were still out. Al is a tall, powerful man and probably the best Rhode Island charter captain for yellowfin and bluefin tuna. He makes his base in Point Judith, Rhode Island. He has written a best-selling book, *Bluefin Tuna*, and has tagged and released over three thousand tunas as part of a NMFS scientific program. Thirty of the fish he tagged have been recaptured in the Eastern Atlantic and in the Bay of Biscay off northern Spain. In the year 2000, 80 percent of the tag recoveries reported to NMFS have been from Al's tags. These recovered tags have given scientists reason to believe that there may be one stock of bluefin tuna instead of the present theory of two distinct stocks. The issue is still being hotly debated at ICCAT.

Al also has a nickname, "One-minute Al." I have sailed with Al many times. It is a three-hour trip to the tuna grounds. The first hour of the trip is hyped-up fishing talk with the mate or on the bridge with Al, made possible by the large amounts of caffeine consumed in the wee hours of the morning, generally before dawn. The second hour is spent trying to snatch a nap curled up in some corner of his boat, and the third hour, restless, in fitful anticipation of reaching the grounds and hooking a tuna. Once Al reaches his destination, confirmed by both his sense of the water, the cavorting bird life, coordinated with the numbers on his Loran machine and the activity on his color depth-recorder, he does not stop the boat and just begin fishing. He wanders among the fleet, sometimes numbering a hundred recreational and charter boats, observing both the other vessels and his equipment, looking for what he calls, "signs." When he finally stops, most often than not, within one minute of putting out the baits, the angler gets a strike. I have seen him do this time after time. The nickname is well earned.

Not twelve slips down from Al's boat that morning, Captain Davey Preble went through much the same routine on his charter boat *Early Bird*. Dave is a schoolteacher by profession, but fishing is his passion. He exercises this passion by taking others fishing on his charter boat. Dave is president of the Rhode Island Charter Boat Association, and he is also serious about conserving marine resources and has served on a multitude of state and Federal fishery committees. Dave called Al the night before and shared the Loran numbers, 14400/43600 he received from the commercial draggers. This would put both of them at the southwest corner of the dumping grounds, also called The Dump, because the Navy had sent to the bottom thousands of unexploded ordnance during and after the First World War. The Dump is fifty miles from Montauk Point, Long Island, and about forty miles from Point Judith, Rhode Island. It was an easy day's sail for charter boats and the private owners of the sportfishing fleet.

When Al and Dave arrived on the fishing grounds, bird life had run riot. Shearwaters were gliding everywhere looking for scraps and the stormy petrels, called tuna birds, were bobbing and hopping from one small wave crest to another issuing their piping cries over the oily slicks that were omnipresent. Fishermen notice these tuna signs and know that below the ocean's surface marauding gangs of yellowfins were slashing and devouring their way through shimmering, massive schools of squid, sand lances, halfbeaks, and butterfish. The cry of "fish on" echoed through the fleet, which at ten o'clock numbered about twenty boats. A plane, an old high-wing Stetson, was making ninety-degree two-mile-square turns over the fleet.

On the horizon the silhouette of a much larger vessel became visible. It was the *Connie Jean*, a purse seiner of one hundred and fifty feet in length; she was the newest of five purse seine vessels that had special permits to fish the Western Atlantic in the United States's declared exclusive economic zone (EEZ), that water within two hundred miles of the American coastline. Three of the vessels, *Connie Jean, A.A. Ferrante*, and the *Eileen Marie*, were owned by Leonard Engrande who lives in Saint Petersburg, Florida; one, *White Dove*, was owned by Mike Genovese and his family from Cape May, New Jersey, and the fifth, a mystery ship, *Ruth-Pat*, was purportedly owned by Sonny Avila from New England. The *Connie Jean, A.A. Ferrante*, and *Eileen Marie*, prior to 1973, had been commanded by California authorities to leave the area because of their destructive fishing practices. The method of purse seining is simple. A spotter plane finds a school of fish, contacts the purse seine vessel, relays the type of fish in a school and its size in metric tons; bluefin or yellowfin tunas, mackerel or herring. The vessel then encircles the school with a large net pulled around the school by a smaller boat, usually thirty feet in length, that is dispatched from the larger purse seine vessel and attaches a line where the two ends of the net meet. This smaller boat begins motoring away from the net at a ninety-degree angle; the drawstring of the purse being attached to a cleat on the stern. The circumference of the net becomes smaller as the purse is drawn tighter. A hydraulic power head off a block and tackle mounted on a boom of the large seiner helps haul in the net and its contents. The size of an entrapped school could be a hundred metric tons and handled easily by the ship's equipment. The entire contents are brought to the side of the purse seine vessel. In the case of bluefin and yellowfin tuna, almost all are dead or dying beyond recovery because the fish's physiology requires it to swim to live. A tuna will die rapidly if it cannot pass water through its gills. The catch is thus killed and hoisted aboard and put into the hold of the ship. The containing areas are well refrigerated by ice and a slurry brine freezing mixture. Marine fishery managers claim to like this method of commercial fishing because the targeted fish are usually caught, thereby eliminating by-catch problems, and accurate counts can be obtained, with limited monitoring and observer access. Reducing costs of management leads the fishery managers at NMFS to wax ecstatic about purse seining, calling it a clean fishery.

There have been many unconfirmed reports over the years that the Kennedy clan has an interest in the fifth purse seiner, the mystery ship *Ruth-Pat*. She rarely sails from port to fish for tunas, but she shares equally with the other four purse seine vessels in any quota and proceeds from catches. Several investigations have always reached a dead end at the Coast Guard documentation bureau. The politics of grandfathering these five permits to the five purse seine vessels have not been lost on NMFS and its lobbyist, the powerful East Coast Tuna Association that manages to obtain letters of support for whatever they profess from Senators Kerry and Kennedy from Massachusetts, Snow from Maine, and Stevens from Alaska, Congressman Barney Frank from Massachusetts, W. J. (Billy) Tauzin from Louisiana, Don Young from Alaska, and all of the other members of Congress who hail from districts dealing with the commercial fish-landing seaports from Maine to Texas, from Baja California to Seattle, Washington. Their present bias emanated from the offices of Gerry Studds also from Massachusetts who knew a thing about politics, for it was his ancestor Elbridge Gerry for whom the political word "gerry-mander" was coined. Gerry and the present aforementioned members of Congress never issued a letter or put forth legisla-tion that did not call for more killing of our fishery resources. When Gerry Studds retired, he received a bit of political pay-back by being the recipient of the Atlantic States Management Fishery Council's conservation award. I found it difficult to take that organization seriously ever again.

Of the 1,450 metric tons of the American bluefin tuna quota for that year of 1993, 475 of them representing one-third of the United States quota were doled out to the five purse seine ves-sels. These vessels employ a crew of eight each, making a total of forty jobs for the five vessels. Their right to the bluefin tuna fishery was protected by the vessel permits issued by our gov-ernment. That so few people and vessels got such a high per-centage of the quota always seemed to be an inequity to the rest of the users of this public resource. None of the owners paid anything more than a token fee to the government to obtain these valuable permits. No auction was ever held. The cost of these permits was political contacts. The purse seiner owners I have met admit they give money politically. I was told that a summit meeting with Ron Brown of the Clinton era cost two hundred and fifty thousand dollars. Oh, to have been a fly on the wall for that meeting. Remember that bluefin tuna roam the high seas at will. They belong to no one group or country. Fishery managers call them a highly migratory species, or an HMS. They are caught inside America's two-hundred-mile zone fashioned by the first passage of the Magnuson Act and as amended currently, now called the Magnuson-Stevens

Sustainable Fisheries Act. If I wanted to invest in a sixth purse seiner, I could not then nor now obtain a permit to build and fish such a vessel and participate in United States bluefin quota. No due process proceedings for granting these permits were ever held. I believe that they are a special fiefdom or cartel granted for some previously undisclosed, perhaps sealed forever, political deal. The tentacles from that deal are long and strong. Whatever Leonard Engrande, owner of the *Connie Jean, Eileen Marie,* and the *A.A. Ferrante* wanted, he usually got. That was until the morning of August 11, 1993.

The *Connie Jean* had no trouble on that day locating the red-hot tuna action. The plane, the boats, and the birds were weaving their tapestry three miles and three hundred degrees ahead. The following is a conversation told to me by one of the pilots as he remembers it.

Plane:	I got a large school spotted. I would say about fifty tons.
Connie Jean:	Good, we will be there in ten minutes.
Plane:	It's crowded so be sure you have room to make a set.
Connie Jean:	We will be fine. Are they yellowfins? The bluefin season doesn't open for three more days.
Plane:	Yup, yellowfins they are.
Connie Jean:	We have company this morning. Pat Gerrior, head of the observer program at NMFS, is aboard.
Plane:	Great. What is the latest price for yellowfins?

Connie Jean: About three dollars a pound. How big is that school? Fifty metric tons? Well, that's over a $300,000 set; not bad, but I tell you what. If they're bluefins, they would bring five times that or fifteen dollars a pound, or about $1,600,000, but the bluefin season is not open as yet.

Plane: You know who to call. Concentrate now. Five boat lengths, about five degrees to the right should do it. Keep coming, keep coming, now.

The *Connie Jean* stopped and the catcher boat was dispatched and began to circle the school of tunas located by the spotter plane. The next hour was recorded on several videotapes by the local charter and sports fleet. As soon as the *Connie Jean* pulled the net next to her port side and began to lift it, several very familiar shapes appeared in the net. It was immediately apparent to all observers nearby, including Al Anderson and Davey Preble, that the net contained not yellowfins, but very large bluefins, and the catch was illegal because their season had not yet opened. The VHF radio began crackling with chatter.

Did you see the size of that giant bluefin? Must have gone seven or eight hundred pounds.

Jesus, she's got seven big bluefins on the deck so far.

What the hell is that captain doing? He should open the net and let them go.

Al, Davey, get someone from NMFS on the phone. Those tunas will all die if he holds them in the net another ten minutes. They all stopped swimming.

Christl! It's forty-five minutes now. They will all be dead. That
bastard. Cut the net! Cut the net!
Someone call the Coast Guard. Report the *Connie Jean* and
give them the Loran numbers 14400/43600.

The tunas were held in the net for over one and one-half
hours. Finally, someone opened the net and the entire catch
went to the bottom. The sports and charter boats edged closer
to the *Connie Jean* despite a NMFS rule that states a distance
of one hundred yards must be maintained, and began taking
photographs and videotapes of the carnage. Several boats
reported a massive red blob on their color depth recorders sink-
ing slowly to the bottom. It was obvious: no tunas escaped the
slaughter. Effectively, the *Connie Jean* and her owner tried to
take an illegal catch, could not, and sent the entire bluefin
school to the bottom for crab food.

The next day, my phones rang off the hook with descrip-
tions of what had happened. I called Dick Stone, head of highly
migratory species at NMFS, and asked him for an explanation.
Dick told me to call Pat Gerrior, NMFS's head observer and the
person in charge of the program for the New England area. She
just happened to be on the *Connie Jean* during the incident. In
fact, later photographs revealed that she was looking straight
into the net as it was brought alongside the vessel. I asked Pat,
"What happened?" She stated that the "crew was unorganized
and confused, and they were astonished they were bluefins
instead of the smaller yellowfins." I then asked if the captain
made any phone calls requesting to keep the entire catch. She
said that several calls were made, but they could not get the

right party to make a decision. I then asked if the seven bluefins that were observed on the desk of the *Connie Jean* were illegally taken to port and she stated that they were not thrown overboard, but was not sure of their final disposition. She did not remember the net's being in the water for over an hour, but she did admit that, if it were, it would have killed the entire catch. What really happened is that the captain tried to get permission to land those bluefins four days before the season opened. They would have been worth twice as much as normal for bluefins because no Western Atlantic tunas were on the market. Permission could not be obtained, however, and Gerrior should have ordered the net to be opened immediately. If done within the first thirty minutes, fifty tons of giant bluefins could have been set free and saved.

In response to the many phone calls I received and my subsequent conversations, I wrote an article, "The *Connie Jean* Incident," which appeared in the *Edge Big Game Fishing Report* for the Fall 1993 issue. This two-page piece is accompanied by two one-page broadsides that I did for the Fisheries Defense Fund, shown at Exhibit B on pages 188 and 189.

Al Anderson was one of those fishermen who called my office on August 12. After describing what happened, he said that he was writing an article for *Fisherman*, a weekly magazine published by the Reina family of Long Island. Al said he had plenty of support from many who witnessed the carnage and they would be willing to testify at any hearing or court action. What is so ironic about the netting of the bluefin tunas is that I have heard the pilots testify in public hearings at NMFS headquarters about the accuracy of their work. "We can identify a gnat's eyebrow," was one statement. *Fisherman* published Al's article in the very next edition. Within two weeks of publication, Engrande sued *Fisherman*, Al Anderson, and me for libel, claiming that we ruined his reputation. Typically this is called a SLAPP suit or muzzle suit.

> SLAPP (slap). *Abbr.* A strategic lawsuit against public participation—that is, a suit brought by a developer, corporate executive, or elected official to stifle those who protest against some type of high-dollar initiative or who take an adverse position on a public-interest issue (often involving the environment). Also termed SLAPP suit.

Engrande wanted to keep us busy defending his action against us before we could gain momentum for a hearing that might have revoked his license or subjected him to a heavy fine or both. I hired the *Boston Globe*'s libel attorney. I calculated that the *Boston Globe*, sitting in the middle of Kennedy territory, must have had vast experience in defending libel suits. I was correct. At our first meeting, the firm's lawyer assigned to the case said, "Mr. Sloan, you just lucked out. If a person sues for libel and ruination of his reputation, you, as the defendant, have a right to every document known to mankind that is in his files." "Does that include all his payroll records, tax records, bills of lading, and air freight shipments to Japan and his original ship's log?" I asked. "Absolutely, and, by the way, Engrande is taking this very seriously; he hired Paul Tsongas's law firm. Paul himself has taken an interest in the case." Pretty high-powered counsel for a fishing vessel case, because Tsongas was a Democratic senator from Massachusetts who took a run for the presidency during the 1992 primaries, losing to Clinton. I told our lawyers to serve Engrande with a subpoena to obtain all his records. I knew he could not stand such scrutiny. Many payments in the fish business are in "green" or "cash-cash" as it is called. I reasoned that Engrande was not above skimming or non-reporting practices. In any case, his documents would verify this. The entire worldwide fishing business is based on the principle of two or three sets of books; vessels with false bottoms hiding extra or illegal catches from prying eyes of port inspectors, a common feature aboard Spanish commercial vessels; and dockside payments usually made in cash because the fish are sold for cash all the way up to the consumer, who may pay with a credit card once in a while. It is a worldwide, gigantic money-laundering problem and tax evasion scheme. It makes the buyout of fishing vessels from an overstressed fishery so difficult because the fishermen cannot replace the "green." It would be naive to think that greedy politicians from all countries are not in on the game. No one talks about it, but these under-the-table payments are pervasive in the world's fishery markets.

Engrande got the message from our motion demanding the production of all of his papers. His lawyers kept trying to exclude almost everything on our subpoena list, hoping that we would give up and disappear. We did not. Here was a golden opportunity to bring to justice a great modern-day sea pirate. In fact, in the spring of 1994, in the middle of our lawsuit, the *Connie Jean* appeared off Pinas Bay, Panama, purse seining for tunas in a governmentally protected marine area. The Pinas Bay Reef is a stalagmite-like rock formation that rises off the ocean floor, is thirty feet wide and two hundred feet high. It is located six miles north of Pinas Bay Lodge, one of the greatest billfish sportfishing areas in the world. The lodge is run by Terry Kitteridge, a fellow International Game Fish Association (IGFA) trustee, who bought the property from Edwin Kennedy, Jr., the oil partner at Lehman Brothers who sent me down to evaluate the real estate for a potential sale. I was president of Lehman Realty Corporation at the time. Edwin and Eddie, his son, were good fishermen and friends. Terry received a report over the VHF radio that a purse seiner called the *Connie Jean* had made a set with her net and already had some bigeye tunas on deck. Terry called the authorities in Panama City, raced down to the dock and jumped on one of the fleet's thirty-one-foot Bertrams to confront the *Connie Jean*. Within thirty minutes Terry pulled alongside the vessel and asked the captain what he was doing fishing in protected waters. The captain waved a piece of paper in the air and said that he had been granted permission to fish. Terry requested permission to go aboard to inspect the document, which was denied. The captain said that he would come aboard Terry's boat, being most anxious and careful not to let her see what fish were on the deck or below. In about fifteen minutes the captain was aboard Terry's Bertram. Terry looked at the purported permission document, which turned out to be a receipt from the Panama Canal Authority for tolls and registration for passage. It was certainly not a document allowing commercial fishing in a protected marine area. The VHF on Terry's Bertram crackled with the news that the government was dispatching a seaplane to investigate the allegations. The *Connie Jean*'s captain beat a hasty retreat and began to go south at full throttle toward the Colombia border only twelve miles away. The Panamanian plane arrived too late to conduct an investigation. The *Connie Jean* never showed up at Pinas Bay again.

After filing numerous cross motions in an effort to avoid having to produce the evidence requested by our attorneys in defense of the Engrande libel suit, we heard from the Tsongas law firm about settlement discussions. They asked what it would take to settle the matter. I knew that this wily pirate would never agree to pay me any money directly because it would be an admission of his guilt and word would get around NMFS like wildfire that he had been defeated. I gave instructions that I would settle the case if Engrande would make out a check to a charity acceptable to me. Engrande wanted the settlement money to go to some charity in Gloucester, Massachusetts, a hotbed of commercial fishing interests. I rejected that idea and he begrudgingly donated ten thousand dollars to the Helen Keller Fishing Club for the Blind, in Brooklyn, New York. I had given several talks at the club where I found great enthusiasm for the sport of fishing. I settled the case because many who promised to give financial aid, or to testify, or to sign affidavits failed to do so because they were physically threatened and were told that their boats would be burned to the water line. This was a shame because we were within fifty thousand dollars of getting all of the information needed to put an end to illegal purse seining and other violations that occur all too often. My decision to settle was also due to the *Fisherman* magazine's actions. It had published Al Anderson's article that Engrande claimed had libeled him; it was codefendant with Anderson and me; it had libel insurance, and yet it immediately bailed out of the case by just exchanging releases. The magazine left its published author, Al Anderson, and me hanging for the costs of defending the suit. I told Al that I would protect him in the case, but the financial burden, carrying it alone, was not fair, nor proper. The Helen Keller Fishing Club for the Blind was delighted with the gift. So was I.

As I sat years later, in the United States Delegation room at the 2001 ICCAT plenary session in Murcia, Spain, I recalled the events of 1992-93 with the *Connie Jean*, Leonard Engrande, and the purse seining industry. Nothing seems to have changed. Fifty percent of the bluefin tuna catch in the Mediterranean are fish under five pounds in weight. A great percentage of that catch, 17,500 metric tons, are zero-age class fish weighing less than one pound. They fit into a medium-size jar, are all caught by purse seining methods, and are sold like sardines through France, Italy, Spain, and Portugal. We were taken to a giant bluefin tuna pen off Alacante, Spain. Purse seiners now entrap the entire school, tow the giant tunas in the net at a speed of one knot, sometimes for hundreds of miles, and put the entire catch in a tuna trap. Once in the trap, the tunas are fed, fattened up, much like a feed lot for cattle, and then are harvested on demand from the Japanese buyers. Japanese money is behind the traps, distribution, and selling. One trap alone sold 1,900 metric tons of bluefin tuna during 2000. The value of that catch/sale was over one hundred million dollars. In addition, 50 percent of the purse seine catch in the Gulf of Guinea off Africa, including yellowfin, bigeye, skipjack tuna, other tuna-like species, and albacore are fish under three pounds. This West African fishery involves more than one hundred purse seine vessels. All of the aforementioned species had been termed fully exploited or overexploited, according to the scientists at ICCAT.

Man's ability to overexploit the earth's natural marine resources remains at high and unacceptable levels despite the management efforts at ICCAT. We need science but, without compliance, the effort is for naught; this leads to defiance. I mused that the bad news is that we have learned nothing; the good news is that we keep trying.

CHAPTER 5

Office of Winthrop Rockefeller
Little Rock, Arkansas
May 1994
1300 hours

● ●

Win Rockefeller, son of Winthrop Rockefeller; Tim Choate; and I had started The Billfish Foundation in 1990. Win had approached me at an International Game Fish Association meeting in Miami where we discussed forming an organization whose sole purpose was to protect the billfishes of the world from overexploitation. I believe that he got the idea from Tim Choate, a lawyer from Miami, who felt much more comfortable running his fantastic fishing camp in Guatemala than running the office of our new foundation. In any case, I agreed to serve as vice-chairman with Win as chairman, and Tim as president and CEO. In 1990, we added a few more trustees: Don Tyson, "Mr. Chicken" from Arkansas; Agie Vicente from Puerto Rico; Paco

Rangel from Mexico; and Don Stott from New York; with fellow IGFA trustees Jack Willits from New Jersey and the Bahamas and Peter Fithian from Hawaii rounding out the board.

Win is a charming man, well-educated, fluent in Spanish and French, with a sly self-effacing humor. He is, after all, the only son of Winthrop Rockefeller, former governor of Arkansas and one of the famous five Rockefeller brothers. I will not attempt here to rewrite the history of the Rockefeller family, but one cannot be part of a Rockefeller endeavor, albeit the tiny Billfish Foundation, without some sense of the power of the family. I am convinced that Win had real concern for the plight of billfish on the high seas because they were being slaughtered by the longline fleets around the world, but it was also in his interest to do something about the problem.

At that time, Win owned the largest Hatteras fishing machine dealership in the South and was to buy three more before selling all in a fire sale. He was a serious billfisherman and had a business that was closely interwoven with the conservation of billfish stocks. Who would buy a $500,000 to $1 million Hatteras if the oceans were void of game? From its inception in 1990 to 1994, the foundation careened from one weekly financial crisis to another. Don Tyson continually, with the help from me and some others, kept the sheriff from the front door and the payroll checks from not bouncing. Oddly, Win hired all of the CEOs after Tim Choate left for his less aggravating and less demanding fish camp in Guatemala. The books were always in chaos. The offices that Win provided free of charge in one of the Hatteras dealerships seemed always in a state of nervous upheaval. Mailings were always late, membership lists were lost, and bills for dues were never sent out or were months late. No one ever understood the lessons of an even cash flow. Worst of all, promises made at one board meeting were always miscalculated and, at the next board meeting, apologies kept coming. Win sent his own family accountant and lawyer in periodically to help straighten out the ineptitude, but even they succumbed to the disorder and their reports were a jumble of incomprehensible assumptions strung together on the basis of erroneous information.

The only saving grace was the passage of a governmental order from NMFS that billfish such as Atlantic white and blue marlin as well as sailfish and spearfish were not for sale. Even possession was illegal. If caught on longlines they were to be thrown overboard dead or alive. Recreationally, the sportfishermen had already self-imposed rules since the 1960s that resulted in billfish releases for fun, and tournament fishing that was almost a hundred percent efficient. Sportfishermen were allowed to bring in a mount for their wall, but this also became passé because the taxidermy industry began to use fiberglass molds for the mounted fish. All the successful angler had to do was measure the length and girth of his fish. He could then put the fish back in the water, release it, and the taxidermist would send him a beautiful replica of his catch. In addition, these new mounts had an indefinite wall life, whereas the old mounts made from the skin of the fish lost their colors and became very brittle after about five years.

In 1994, Win called a special meeting of the executive committee whose members were Win, Don Tyson, and me. I flew to Little Rock for lunch.

Win is a man of great charm. He would like to be a warm and fuzzy friend, but his upbringing and guarded existence make this difficult. He has a room in his house filled with electronic equipment that keeps track of his movements twenty-four hours a day through a device that he constantly wears. His location shows itself on a map in this war room. His pilot, in those days, also served as his bodyguard. At one session of the foundation, Win hosted all of the directors at Win-Rock Farm, one of the largest animal sperm banks in the world. The first night, Win also hosted a dinner for fifty southern district FBI agents and their wives. He also monitored the local police channel and would go racing out to inspect any police call that struck his fancy. In any case, we met for lunch in Little Rock about the latest crisis.

Don Tyson is a no-nonsense man. He makes up his mind quickly and acts, most often with his checkbook in hand. I remember at one board meeting we needed some money. Quickly, Don threw a check on the table for $50,000 and asked, "Do I have two minutes?" "Do you want to say something?" I questioned, expecting a speech to which I would have been delighted to listen. "No," he said, "to get out of here; my plane is waiting at the airport."

After reviewing the recent management calamities, Don turned to Win and said, "Win, let's stop f- - - - - - around with the penny-ante stuff. I'll put up $250,000, you put up $250,000, and Steve will raise another $250,000 from the rest of the trustees." The words weren't off his lips before I began nodding my head, yes. "I can do that," I said as I immediately mentally ran down the list of the rest of the trustees and calculated their contributions as well as my own. Finally we were getting somewhere, I thought. Win's reply left me dumbfounded. It was a simple, "No, I am not going to do that."

41

Don expressed his regrets over having to leave for the perennial airport plane waiting for him. Win drove me to the airport to catch my plane. Not much was said during the drive. I kept thinking that a golden opportunity had just passed by. I had heard that the Rockefellers used their name a lot, but kept a tight rein on their money. This refusal of Tyson's offer, surely, with Win as chairman, with his name all over The Billfish Foundation, was a miscalculation of the first order. Silently, mentally, I put on my IGFA trustee's hat and mused about what good IGFA could do with $250,000 in the till. As I stared out the window, I remembered making a calculation that in three of the past four years I had donated more money to the foundation than Win had, to say nothing of the introductions and the board members that I had brought aboard. Win and I shook hands at the airport. The trip had been a disaster, but there was still more ahead.

CHAPTER 6

Office of Stephen Sloan

230 Park Avenue, New York, New York

Telephone conversation with John Spence,

Acting CEO of The Billfish Foundation

May 2, 1996

0930 hours

• •

John Spence was the latest in a line of managers hired to run the foundation. He was the most capable to that point and, for the very first time, we had someone with us who had his hand on the tiller and knew his course. He called me and said, "Mr. Sloan, this is John Spence at the foundation. I've got a problem. You know that Win [Rockefeller] had to take some time off for a personal family problem with one of his children. Well, I have a check here for $40,000 due to be sent to ICCAT. This is our annual contribution to the billfish observer program in the Caribbean. What should I do?"

I replied, "John, have we ever gotten a written report, photographs, or a videotape regarding this grant to ICCAT? Have we ever received an acknowledgement from any official? After all, it is the only private grant in ICCAT's history."

"No," he answered, "not that I can remember; in fact, the file is empty in that regard."

"John," I told him, "hold on to that check until you hear from me. If you send it out I am coming down there to crack you with a 12/0 tuna outfit." He laughed, and said, "Yes, sir."

I immediately called Dick Stone, head of the highly migratory species division at NMFS headquarters in Silver Spring.

"Dick, I am going to honor the pledge The Billfish Foundation made to ICCAT with the following understanding: first, I want a written report on the observer program in English. I want the log of the trips to date including the one we are just paying for with the check. We have been contributing $40,000 a year for the last four years of work. We do not even have a record of anything for the money we spent. Second, I want a film of the next voyage, which will begin in a month."

"Steve, absolutely no problem," he said. "I will notify Eric Prince immediately. You are certainly entitled to a full report after giving money to us for four years [$160,000]. I will see to it."

Dr. Eric Prince was a full-time employee of NMFS, based in the southeastern headquarters on Virginia Key, Miami, Florida. In fact, he was the only employee who had anything to do with billfish. He sat in on most of the foundation board meetings as a special adviser. Why he did not get a film and the reports I had just asked for on his own initiative was a mystery to me, but now he had orders to do so. It was also Eric who told me that in

Trinidad there was a large thirty-thousand-square-foot ware-house chock-full, from floor to ceiling, with billfish plugs. A plug is the body of the fish, eviscerated, without the head and tail, as shown in Exhibit C at page 190.

Eric was also the person who tipped me off that the Japanese had made a joint venture with Grenada to supply thir-teen longline vessels. He said that the kill of Atlantic sailfish there was massive. In spite of these "tips," Eric seemed content to go to billfish tournaments and count the otolitas, which are the ears from billfish heads, as if they would lead to some great scientific conclusion one day. Eric was against any aquaculture research that might put back these billfish after they spawned in captivity. He based his theory on Darwin's natural selection law, by which only the strongest survive. But Eric also gave me the feeling that he was against these aquaculture programs because they represented a threat to his job. He once gave me a speech at one of the foundation's board meetings in which he said that some new science had shown that if a billfish could be injected with tetracycline, tagged, released, and then caught again, the tetracycline would create a halo-like image around the vertebrae and a scientist could count the rings and deter-mine the rate of growth at sea from its first capture to the sec-ond one. This information was far more important to stock assessments of any species than the present methods of tagging and recoveries, which only showed how many days that fish was at sea and how far he traveled.

I immediately saw that the tetracycline data could have dra-matic impact on billfish stock assessments. It was commonly thought that a sailfish only lived for five or six years in the wild.

If new data proved that they lived for ten or twelve years, then the stock assessments would have to be cut in half. I asked Eric how long he had known about this procedure, and he answered that it had been for perhaps five or six years. When I inquired how many fish he had tagged so far, he replied that he had tagged none.

"I don't believe this," I said. "Do you know how many billfish the men in this room must have released in the past five or six years? It is in the thousands. Many here own a boat, me included, and I would be delighted to have participated in such an important project."

"Well," Eric said, "You have to take into consideration the possible health effects on someone eating a sailfish that might OD on tetracycline."

"That is preposterous," I yelled. "Don [Tyson], tell him what you feed your chickens when they hatch and thereafter."

"Tetracycline," Don said emphatically.

I was really frustrated with his resistance and said, "Eric! I am going to be in Isla Mujeres, Mexico, in late April and I am going to tag at least ten sailfish and inject them with tetracycline and put a special tag in that fish."

"Can't be done," challenged Eric.

The gauntlet was thrown and the challenge accepted.

That April, with Mexican and American scientists onboard, we tagged ten sailfish that I hooked and brought to the boat. All were returned to the ocean and swam away with the special tags streaming from their shoulders. One was recovered, but the fisherman from Key West who returned the special tag unfortunately smoked the fish and ate it. Now we have special sophisticated electronic tags that pop up on demand and release data to satellites circling in the heavens. The new electronic tags will have special meaning, as we shall see in a subsequent chapter on bluefin tunas. I still feel disappointed that NMFS staff had known about the tetracycline tagging and its importance and did nothing for five or six critical years regarding this important advance in science on billfish assessments. We could have started a program on white and blue marlin as well.

About ninety days after John Spence's phone call and the release of the check, I received two hours of videotape filmed aboard a Venezuelan longline vessel fishing near La Guaira, the seaport of Caracas. The ship's deck was a killing field worthy of the Khmer Rouge. During the same period of time, I was visiting Murray Brothers tackle shop in Riviera Beach, Florida, when a salty captain came up to me and asked if I wanted a tape about longlining in the Gulf of Mexico. I went to his house and got another two hours of tape that was filmed on a trip made in the Gulf. This film graphically showed how swordfish and bluefins are butchered at sea. Simple; all that was needed was a hacksaw and ten seconds of time. In the film, fish were still flopping on the deck even with heads cut off. What was most important about that Gulf of Mexico film was that it showed that the crew was high-grading. Under the law, a longliner was only allowed one bluefin tuna per trip. In the film, the first tuna that came aboard was about four hundred pounds and it was butchered almost immediately. The second that came aboard was about seven hundred pounds. It was supposed to go overboard, alive if possible, because the vessel had already killed one tuna. This crew did not follow the rule. They landed and kept five bluefins.

Both films were devastating in their impact. Both clearly, con-
clusively showed that billfish—the marlins, sailfish, and also
the sharks—were very much alive next to the boat and could
have been released with little effort on the part of the crew.
Nothing was released. Everything was taken aboard, including
rays, skates, small tunas, baby swordfish, lance fish, and sharks
including a pup mako of ten pounds. Endangered loggerhead
and trunk-back turtles were hooked with regularity. The films
were a gruesome reminder of indiscriminate killing at sea by the
longliners. The prime target in both cases was swordfish, but the
by-catch was nine to one over the swordfish. I began to wonder
if the oceans could take such a slaughter on a daily basis.

CHAPTER 7

The Billfish Foundation Board Meeting
The Sailfish Club
Palm Beach, Florida
November 1996
1200 hours

• •

The meeting was in session and item three on the agenda called for the review of the board's position on the longline film that I had obtained from Eric Prince. Mark Sosin, a good friend and noted sportfishing writer and television personality who hosted his own excellent show on TNN every Saturday morning, opened the discussion by showing a shortened version of the two-hour longline film. Mark had narrated it and, while it looked pretty polished to me, he said that it needed some more editing and perhaps some better music in the background. I wondered with some amusement if he meant the funeral march of Mendelssohn. There was a lively discussion after the film was shown. Basically, we were divided into two opposing camps. Almost every trustee in the room favored getting the film into some public awareness program and on television. The carnage shown in the film, we thought, surely would get the public behind any movement to curtail this destructive type of fishing, and enhance our efforts to get all longliners, domestic and foreign, to release billfish at the boat. The film clearly proved that most billfish are alive when pulled alongside the fishing vessel.

The other view was held by Win, who argued against releasing the film until we are fully sure there will be no repercussions. Win was holding something back. To this day in 2002, the film has never been released for television, despite increased carnage by the longline fleets and their devastation of billfish stocks. He did not fully express himself. I could only infer that the Rockefeller family had long and deep political tentacles into South America, especially Venezuela where the film was taken, and did not want any repercussions from those countries if the film were received with anger.

The board, following Win's lead, voted not to show the film on public television, and to try to work on getting a more polished, ready-to-view version, despite my heated pleas for its immediate release. No money was voted for the project and Win did not solicit any from the trustees. I might add that Don Tyson was absent from the meeting so the chance of another immediate funding was dim without him. In boardroom politics this is known as a muzzle job. I shook my head in amazement. Here we had bona fide evidence, produced with our own hard-earned $40,000-a-year contribution to ICCAT. Billfish were being brought to a longline boat alive time after time, and we, the leaders of the billfish conservation movement, refused to release the facts. I left the meeting more determined than ever to introduce Don Tyson to the IGFA board of trustees. I called Don and asked him the consummate question: If asked, would he serve? He said he would be delighted. I proposed him at the next IGFA board meeting a week later. Today, thanks largely to Don Tyson, IGFA has a fantastic $30 million interactive museum, a fishing hall of fame, and a twenty-thousand-book library, next to another IGFA trustee's business, Johnny Morris's Bass Pro operation, in Dania Beach, Florida. This outstanding complex is located off Interstate 95 at the Griffin Road Exit, one mile south of the Fort Lauderdale airport, and the foundation is still careening from one crisis to the next, and has yet to release the film for public consumption.

In an almost bewildering set of events, in a guilt-ridden moment of catharsis, Ellen Peel of TBF secretly signed the famous Memorandum Of Understanding (MOU) with the arch-enemy longliners, the villains in the still-repressed film. Recently, four longtime board members quit because they were not informed about the memorandum of understanding and the suppression of the film. Its former chairman, Mel Immergut, a New York lawyer who was head of Milbank Tweed, a prestigious law firm, had to resign as a board member because he sold bluefin tunas worth thousands of dollars without the proper permit, to say nothing of the image of protecting the bluefin tunas from exploitation. But I am getting a little ahead of my Kafkaesque story of the foundation's continuing trials.

CHAPTER 8

Maine Fishermen's Forum
Samoset Resort
Rockport, Maine
March 1997
1000 hours

• •

Each year the Maine Fishermen's Forum holds a conclave at the Samoset Resort in Rockport, Maine. I had been invited back to give the "recreational" speech to about four hundred serious commercial fishermen, whose opinion on sportfishermen could range from "rich spoiled brats," to utter contempt, to outright hatred. I had prepared my usual commercial cooperation speech, which was laced with such platitudes as: "We must work together," and "No man is an island," and "These are public resources that everyone has a right to use, not just commercial fishermen," and "The economics of sportfishing are as great a value as the commercial landings of forty-nine states." Alaska is the exception because its landings of commercial fish products are as big as the other forty-nine combined. I wove these themes in and around some humorous, to me anyway, fishing anecdotes. It was soon after I arrived that I noticed a quiet, desperate mood at the meeting among these commercial fishermen. NMFS had been publishing rules, combined with several environmental lawsuits, to close much of the fishery in New England because the science had shown that the groundfish stocks had been severely depleted. Tony Bullock, former mayor of New Bedford now at NMFS, was at hand in his new job as coordinator of the commercial industry to NMFS. (Recreational fishermen and their industry had never had a special representative.)

He was a good choice, but clearly uncomfortable about bringing the news of impending closures to his former district voters. Bullock was bringing about $6 million in grant aid from NMFS to the entire New England fishery. The message was that NMFS would try to help them get loans from their friendly New England banker: an oxymoron if I ever heard one. I realized immediately that the fear that I saw in the eyes of these hard-working commercial fishermen was real. Most of them had borrowed on their homes, either directly or in the form of guarantees, to buy and retrofit their fishing vessels. They stood to lose everything if the banks called their boat loans and they would lose everything if their fisheries were closed for any extended period, even if the future might bring the stocks back. I felt that Bullock's aid package, while well-intended, was a Band-Aid trying to cover a massive hemorrhage.

I revised my speech and wrote out a case for Fish Recovery Bonds. See Exhibit D at page 196 for the text. In essence my plan was to create a trust fund much like the highway trust funds that built and maintained the interstate highway system. Everything used at sea including the catch itself would be taxed and that money would be paid into the Fish Recovery Bond Trust Fund, for want of a better name at that time. I calculated that about fifty million dollars a year for the next twenty years could be raised from taxes on anchors, chains, gear, lines, rods, reels, marine fuel, slips, and so forth. That fifty million dollars would be the debt service on five hundred million dollars of FRBs (Fish Recovery Bonds). Here was a fund that could say to a fisherman, "Sorry, but you cannot fish for the next three years. We will pay you to sit on the sidelines until the fishery recovers. If you do not make a living as a carpenter, gardener, painter, or truck driver, we will subsidize you out of the fund. It is a 'win-win' all the way around, without the hardship of losing your boat and, indeed, your house." It was obvious to me that commercial fishermen do not understand that the capitalization of steady cash flows could work in their favor. In my example, the bonds would be self-amortizing over twenty years, exactly the time set by NMFS for many of the fishery management plans for many of the overfished species.

I must say, there is an innate fear among commercial fisherman about stopping their fishing. Many feel that they will never be allowed to fish again and that would be worse than any imposed quota system.

My speech was published in the *Commercial Fisheries News*, and the underlying idea still makes sense today. The idea needs a leader in Congress. One has yet to emerge, and the likely players, Kerry, Kennedy, Snow, Breaux, and Tauzin, are too bogged down in pork barrel politics to take on a program like the one I suggested. Prior to the conference, Dick Stone, head of highly migratory species at NMFS, published a notice in the *Federal Register* that he intended to hold a billfish scoping meeting on the second night of the conference. I thought this to be curious since there had not been one billfish caught in the Gulf of Maine in the previous twenty-five years. I wondered what possible support he was going to get among commercial fishermen for billfishing when the roof seemed to be falling on their collective heads from the failure of all the fishery management plans NMFS tried to foist on their businesses. How naive I was back then.

Scoping meetings are required by law (the Magnuson-Stevens Act) before the NMFS can publish a final rule that becomes the fishery law of the land. These meetings are for NMFS to hear out the public, get the comments, publish a synopsis of those comments, and then publish the rule. Since 1997 was the year of the billfish at ICCAT, scoping meetings for billfish would have some effect on the United States's position at ICCAT. As was customary, a sign-in sheet and a request to speak was available at the table. The meeting was held at nine o'clock in the evening in the middle of a blizzard. I noticed about three hundred people in attendance. All roads to town were closed because of the blizzard, so the billfish scoping became the evening's entertainment. About twenty had asked to speak.

Nelson ("Hammer") Beideman, head of Blue Water Fishermen's Association (BWA), the lobbying group representing the longliners, was first on the list. I was second. Nelson is about five foot five and has wisps of hair across his head. His

eyes are always darting, looking, and alert. Hammer may not
have a college education, but he is plenty smart. He got his nickname for his maniacal insistence on calling the NMFS home
office every day, hammering his points. He is street smart, fish
smart, and plain smart; relentless in protecting his industry
from any real or perceived harm. I have seen him pound away at
some arcane swordfish point of order until it was black and
blue, and then come back for more until NMFS gave in. Many
times this method was very effective. I have seen the personnel
at NMFS cringe knowing that they would have to take him on.
Nelson had plenty of ammunition for his cannon. He claimed,
and most of the time correctly, that all foreign national vessels
and not the American longliners were out of compliance and
taking his beloved swordfish from the Northern Atlantic and
Caribbean waters and then selling (exporting) to the United
States markets. Incredibly, the United States's imports of
swordfish for the past few years have equaled the American
longline quota granted by ICCAT. Clearly this is not a level playing field for the American vessels. The high-powered influence
that Nelson exercises on NMFS is curious because the whole
American fleet consists of two hundred and sixty vessels and
the whole longline industry only employs about a thousand people. The economic value of their catch is from about $6 million
to about $10 million annually. This is not even the volume of a
good Cadillac agency. I do not know any Cadillac dealer who
has a million-dollar-plus federal payroll working for his interests, all on the taxpayers' money. I have often wondered what
political muscle Hammer has over the agency to rate such attention. Many vessels in the fleet are old, tired, wooden, and non-

competitive in worldwide longlining. Nelson managed to negotiate 28 percent of the ICCAT swordfish quota for his American fleet. This approximates 3,000 metric tons.

I should have known something was amiss that evening at the Samoset Resort. It was a good thing I decided to go. The first words out of Nelson's mouth at the scoping meeting were, "The longliners should not be required to release billfish when they hook them as part of their by-catch procedures. The fish are dead anyway and it is a waste of a fine resource. Who are these sports guys, anyway? Why should we not be allowed to retain billfish and bring them to shore and sell them for whatever they are worth? We can develop markets for them." By the time he was finished, I was baited for bear. I had brought the two semi-edited longline films with me. I reached into my bag and took out The Billfish Foundation one. I strode to the podium, my jaw set tightly, and opened with, "Mr. Stone and Mr. Beideman, I have a surprise for you. Nelson just got finished telling this meeting that almost all billfish are dead at the boat. In fact, it was his main argument for a change in the law that would allow his fleet to keep billfish and sell them." I could see the red hackles on Nelson's already red neck. "Watch this!" I yelled. I showed about five minutes of the longline film I had obtained. Fish after fish came to the boat alive and swimming vigorously. The many white marlin, sailfish, dolphin fish, sharks, all were alive, and they only stopped swimming when the long-handled gaff went under their chins and they were pulled alive into the boat to fall prey to the very active machetes of the well-trained crew.

When I turned on the lights, Beideman was purple and sputtering. "Where did you get that film? I am going to sue. That film belongs to the U.S. government, not some sportfisherman. This is illegal, outrageous," and he sputtered into silence. Dick Stone was ashen. I did not let on at that moment that he had obtained the film for me as part of the foundation's annual contribution to ICCAT. Finally, I had struck a blow for the conservation of billfish at a crisis juncture. I began to realize that this out-of-the-way NMFS scoping meeting on billfish was to have been an advance step to a later ruling that might allow the sale of these fish. The scoping meeting closed in rapid fashion. Beideman came over to me and told me that the film would never see the light of day and he would get federal marshals to gather in all prints if he had to. I told Nelson that he was dead wrong, for no judge would ever gag the film if all the facts were known about its acquisition. He muttered as he went away and said, "I'll get that prick, Eric Prince, if it is the last thing I do." To this day, the foundation's film has never seen the inside of a television network viewing room. Several prominent members of the foundation board have resigned over its unseemly burial, including me. In June 2002, I gave a version of the film to the Pew Oceans Commission at a meeting I arranged and hosted with IGFA for their report to the president and Congress. A copy is available for purchase by any reader of this book and a card is enclosed for this purpose.

CHAPTER 9

United States Delegation Room
Hotel Chamartin
Madrid, Spain
November 1997, the Year of the Billfish
2100 hours

· ·

We had just received word from the Japanese delegation that it had rejected our proposal that mortality on longline vessels interacting with white and blue marlin should be cut by 50 percent. Included in the negotiations were suggestions on methods of releasing these marlin as they were brought alongside, hooked on the gear of ICCAT's longline vessels. Not only did the Japanese reject our proposal, they insisted that a reduction of effort be initiated by the United States on its recreational fishermen. This was a curious demand, because the Japanese delegation included a member from the Ministry of International Trade and Industry (MITI), commonly called Japan, Inc. MITI is the most powerful Japanese governmental agency. Nothing moves in or out of the country without its blessing. I guess that it had never heard of the hundreds of millions of dollars worth of goods that Shimano, Daiwa, Yamaha and Honda sell to American recreational fishermen, to say nothing of the Gamagatsu Company, a major hook supplier. Will Martin, head of the United States delegation, merely smiled at Japan's seemingly outrageous request for a reduction of effort on our recreational fishery that was already 95 percent hook-and-release.

Will is a lawyer from Tennessee and was appointed by Vice President Gore. I found him to be calm, reassuring, and brilliant in his thinking. He was also an able negotiator. He had informed us at the beginning of the meeting that this 1997 ICCAT meeting was to be his last because he was going into private law practice. The entire ICCAT process thus had started on a down note because Will Martin was very much respected by all parties in our delegation.

There was much science from the Standing Committee on Research and Statistics (SCRS) and the environmental team's own scientists, reaffirming that unless 50 or greater reduction was put into effect at ICCAT, the entire Atlantic stocks of white and blue marlin would crash. This was the first of the ICCAT species that was not primarily a food fish. The United States's position going into the meeting was to push as hard as possible for a 50 percent reduction on mortality. There was only one way. Get the longline fleets to release these fish alongside the boat. My ICCAT longlining films, as discussed in an earlier chapter, clearly proved that these marlins could be released alive.

The irony was not lost on our delegation. In the United States, over the past twenty years, recreational sportfishermen had released approximately 500,000 billfish. The longline fleet had killed millions and now the Japanese wanted to trade a small reduction in mortality from their vessels (10 percent was the first offer) for a reduction of American recreational effort. Japan has excelled at this kind of hardball politics even in its own country. The fishermen's lobby and political process is the most powerful in its Diet. The government is in lockstep with commercial protocols and objectives. All the politicians are tuned into the process. It is also big business. The Japanese recreational fisherman has little or no say in the process. In fact, my Japanese angling friends tell me they are third-class citizens in their own country as far as marine recreational angling is concerned. Because its own marine waters have been fished out, the Japan Game Fish Association, modeled after IGFA, keeps records mainly of fish caught in distant waters.

I fully understood what the ramifications of the Japanese proposal were and explained it to my fellow delegates. "A reduction in effort means that on Monday, Wednesday, and Friday we will not be allowed to go to sea and fish for marlin or sailfish. No other country in the world would allow such draconian economic measures to be imposed upon a major industry. Try selling boats, gear, or tackle with that overhang. Try selling that to the sixteen million voting saltwater fishermen in the United States. Try explaining that to the self-imposed hook-and-release programs started by sportfishermen in the 1960s for billfish and are now being penalized by the very same countries that have done the killing and therefore decimated the billfish stock. This proposal by Japan is both outrageous, egregious, and purposefully punitive."

Japan got its lead from the story in Chapter Two of this book. The Japanese thought they saw how easy it would be to impose regulations on American recreational fishermen and obfuscate their own killing methods, compliments of the then U.S. commissioner, Carmen Blondin.

Dr. Ellen Peel, CEO and president of The Billfish Foundation, said, "It's a 50 percent billfish mortality reduction or nothing; that's our position." Dave Wilmot, director of the Ocean Wildlife Campaign and delegate from the Green Team's organizations, said in a moment of prophecy (Exhibit E: Ocean Wildlife Campaign News Release, pages 197–198), "If they do not accept the 50 percent, which science demands, we will all be back here in the year 2000, when the billfish matter comes up again, with more of the same problems and arguments." Will Martin, with a deep sigh, said, "We will have to keep punching. I believe that the bottom line is 25 percent and some reduction of effort on U.S. recreational fishermen and other billfishermen from other countries is critical to save these species. In any case, we will keep trying." I handed Will a shortened version of my longline films, reformatted for European VCR use. "Show this to the Japanese delegation if there is any doubt whether these fish could not be released alive. I know you have some doubts about our position, but look at these yourself," I said. Will gave me a quizzical look and took the films.

The next few days provided no surprises. The Japanese, speaking for all of the longline players, came down to 40 percent, then 35 percent, then 25 percent reduction, but still with the proviso that U.S. recreational fishermen cut their effort by the same percentage.

Finally the moment of negotiating truth took place. The Japanese final offer was 25 percent spread over two years with another look in the year 2000, with a commitment from the United States to put measures in place to reduce U.S. landing of billfish. At the delegation meeting someone from NMFS passed a remark about "protecting some rich guys who own boats." I took immediate umbrage at the remark, which was always a well-circulated canard in the NMFS corridors at the Silver Spring headquarters. Commercial fishermen are the poor, downtrodden good guys, while recreational fishermen were rich, fat slobs who deserved and got little or no attention.

"You are mistaking the rich and comfortable for the very rich; there's rich and very rich," and the difference being that the very rich can live handsomely off the interest of their capital and the rich can own a boat and have a good-paying position or own a small business. In most of the rest of the world, only the very rich can own a boat, but in America a person with good credit, a nice business or job can easily find a bank willing to make a boat loan to provide both sport and recreation to millions, creating jobs for many industries. I am sure that NMFS, even as of this writing, is still convinced that only the very rich would be helped by its efforts. It is so hard to change the mindset of a governmental agency. NMFS, which is a part of the Department of Commerce, should be protecting the American sportfishing and boating business and its socio-economic value to this country, instead of ridiculing it. I was serious. There are owners of twenty-five hundred American charter boats called six-packs because of the amount of fishermen they can take on a charter as regulated by the U.S. Coast

Guard. These men have mortgages, bills to be paid, send their children to schools and colleges, pay their taxes, and keep their heads above water. These people are also recreational fishermen. To stay financially alive amid declining fisheries, they deserve the same attention as the fisherman of New England and Beideman's longliner Blue Water Fishermen's Association. "After all," I argued, "these fish are a public resource. They roam the seas at will. They belong to no one and to everyone at the same time. They are part of a public trust, for no one user group has rights over the other. There are quotas and rules, period. I can assure you that the recreational fishermen are treated as second-class citizens by NMFS."

The battle lines were drawn. Dr. Ellen Peel voted to veto the Japanese offer and stick to the 50 percent reduction as suggested by the SCRS scientists. The rest of the delegation was somewhat on the fence. I said, "I, for one, cannot go home and explain to anyone that we allowed fifty thousand more billfish to be killed if we do not get an agreement of some sort from Japan. I am inclined to accept the Japanese offer of 25 percent over two years with another look in the year 2000, providing we can get some protection on the reduction of effort."

Mike Nussman, the recreational commissioner, said, "I agree with Steve, that we are not going to do any better." Glen Delaney, the commercial commissioner, agreed with Nussman. Marion McCall, legal counsel from NMFS, said that she would work with anyone on the drafting of the recommendation for the attention of those at the full plenary session. Marion and Dr. Rebecca Lent, head of the highly migratory species division, and I met to go over the draft.

I looked at clause three and suggested that we add the words, "that each country advise ICCAT annually of management measures in place to achieve the reduction of landings for white and blue marlin by both commercial and recreational interests. If management measures are in place, then no reduction in effort is necessary." This language was to be part of Exhibit F ("Recommendation by ICCAT Regarding Billfish," page 199). Both Marion and Rebecca gave me a hug and said, "This is perfect. The U.S. has management measures in place and therefore it will not be required to place any further restrictions on effort for American recreational billfishermen. Japan will be appeased and we will have a 25 percent reduction over two years. Perfect," they said. One look at document number 71 in Exhibit F, now in its final form, will reveal a glaring deficiency in article three of the document. The last sentence that I had worked out with Marion McCall, "If management measures are in place then no reduction in effort is necessary," was eliminated from the final document. This left the door open to two more years of sniping and grousing by the Japanese over the "outrageous slaughter of billfish by the American recreational community." More cries of "reduction of effort" became the party line from Japan from 1998 to 2000. (In the year 2000, NMFS through its surveys reported that U.S. recreational fishermen killed a grand total of 217 white and blue marlin. This was an infinitesimal amount compared to the 547.5 million baited hooks that are in the Atlantic and Caribbean waters annually catching marlin by the hundreds of thousands.)

In fact, NMFS picked up on the Japanese position. It developed a management philosophy that went according to a formula that I created when I got back to the United States. RE=RM=RMGMT. Reduction in effort equals a reduction in mortality equals a reduction in management. This formula, if applied, gave NMFS the perfect excuse for not committing any further resources (money) to helping recreational fishermen. This was a simple, "Get the alleged perpetrators off the waters and bingo! No more problems. We can save some money and get a raise next year."

Our deliberations in Madrid resembled the Marx Brothers movie when Groucho and Chico begin tearing up a contract piece by piece. It shows what can happen when a document has to be translated into four languages, circulated, snipped, and pieced together again. In the final version of an important document, more often than not, it hits the floor of the ICCAT plenary session during the last two hours of the last day, just before the delegates are heading for the airport. I could have sworn that my language was in the final draft that I received on the last day. Perhaps I was mistaken; perhaps I was not. Unlike treaties, where the Senate of the United States has the last word, this is not the case at ICCAT. In fact, the Secretary of Commerce is bound by what we sign at the ICCAT meetings. It took two years after the ICCAT 1997 Madrid meeting to resolve document number 71.

Even at this writing the Japanese are still insisting on a reduction of effort by American sportfishing vessels. We have resisted, but it is still on Japan's agenda.

CHAPTER 10

**NMFS Headquarters
Silver Spring, Maryland
April 1998
0900 hours**

• •

A recreational fisheries summit meeting was arranged through Terry Garcia, number two official at the National Oceanographic and Atmospheric Administration (NOAA), the parent organization of NMFS, with Dr. Gary Matlock, head of the Sustainable Fisheries Program. Gary held the number two job at NMFS. The approved agenda was to be a discussion of our recreational fishing grievances over the reporting of recreational yellowfin tuna catches to ICCAT. The meeting was approved and confirmed at least three times through Terry Garcia's office and Matlock's secretary. The governmental pecking order here is as follows: At the top is the Department of Commerce, then NOAA operates under it with a budget of $2 billion-plus dollars per year, and under NOAA is NMFS with a budget, as of 2002, close to $700 million. NMFS is the steward of the nation's living marine resources.

Attending the meeting was the organizer, Jim Donofrio, executive director of the Recreational Fishing Alliance (RFA), a New Jersey-based national organization; Mike Nussman, the recreational ICCAT commissioner and vice president of American Sportfishing Association (ASA); Tom Fote from Jersey Coast Anglers Association (JCAA); John Koegler, statistician from JCAA; Phil Kozak from National Fishing Association (NFA), based in New Jersey; Joe McBride, president of the Montauk Boatmen and Captains Association (MBCA); Bob Eakes from North Carolina and a member of the ICCAT advisory panel; Dick Stone, formerly head of highly migratory species at NMFS and now retired and a consultant to RFA; and myself, representing International Game Fish Association (IGFA), the Confederation of the Associations of Atlantic Charter Boats and Captains (CAACC), and the Fisheries Defense Fund, Inc. Many of us got up at four in the morning to make sure that we were on time for this very important meeting. I served on the bigeye tuna, albacore, yellowfin tuna, and skipjack tuna, (known as the BAYS), working committee of the United States ICCAT advisory committee. To be a member of the United States delegation at an ICCAT formal meeting abroad, one has to be appointed by the Department of Commerce, NOAA, and NMFS, with FBI clearance.

Since 1990, NMFS has reported the yellowfin tuna catches by the commercial and recreational sectors to ICCAT as required by the terms of the treaty. When we received the first reported figures in 1990, it was clear that the NMFS position was that no recreational catch could exceed that of a commercial group for that or any other species. The yellowfin catch area is from the Gulf of Maine to the western Gulf of Mexico. In charge of the yellowfin reporting are two scientists from NMFS, both based at the Southeast Fishery Science Center in Virginia Key, Florida, and they are Dr. Joe Powers and Dr. Jerry Scott. Their boss was Dr. Bob Brown who always kept a low, almost invisible, profile and let Powers and Scott take the heat on the many controversies that developed from the stock assessments and quota system. The most ever reported by NMFS for a recreational yellowfin tuna catch was a little over 5,000 metric tons. A spike year happened in 1992 when the commercial catch soared to 8,000 metric tons, 17.6 million pounds. The recreational leadership found out in 1997 that NMFS had told the commercials to get a larger reported catch in 1992 so the scientists could use 1992 as a base year in case quotas might eventually be voted upon at ICCAT. All of us in the room that day to meet Dr. Matlock were fully aware of the peculiar circumstances surrounding the reporting of commercial yellowfin catches, and the deliberate underreporting of recreational catches and landings. Not even considered in the totals was the reality that nearly all yellowfin tuna recreational activity is on hook-and-release basis. While many are hooked, few are killed, and none, under the law, was to be sold in the markets.

The 1998 total Atlantic and Caribbean ICCAT annual reported catch of yellowfin tuna is 650,000 metric tons or 1 billion 430 million pounds. If the United States kept its reports low in the 5,000 to 10,000 metric ton range, eleven million to twenty-two million pounds, this would represent between 3 and 6 percent of the ICCAT-reported yellowfin catches. Therefore, reasoned

NMFS, the United States is such a minor player that ICCAT will impose no quota restrictions and put the United States in a category with minor catches from developing countries and small emerging countries. Part of the strategy for our commissioners was the hope they could negotiate a clause stating that, "Any country whose catches are below six percent of the ICCAT-reported catch per annum does not need to take management measures within its own EEZ." These negative by default schemes always have a way of backfiring, usually, like Murphy's Law, at the worst possible time. One small fact was overlooked by NMFS: In underreporting the U.S. yellowfin catch, we were violating the terms of a treaty that we had signed up to obey.

Unfortunately, as catch levels increased for yellowfin tuna, this "negative approval by default" approach did not come to pass. Each year NMFS would issue a report to our ICCAT advisory committee meetings at NMFS headquarters stating, "Through the best science available, the yellowfin tuna catch for the U.S. both for recreational and commercial users is 5,000 metric tons each." Every year the BAYS committee, at my insistent urging, would meet, and those from the recreational side would strongly voice their disapproval of the reported figures. See Exhibit G at pages 200–204 for texts of BAYS material. The state of North Carolina alone from 1996 to 1998 reported a yearly catch of recreationally caught yellowfin tuna in the 4,500 metric ton range. This one state reported nearly as many yellowfin tuna catches as constituted the entire United States position for recreational sportfishing catches from Maine to Texas.

We knew that NMFS deliberately kept the figures too low. The debates often got heated, angry letters were written. It was like a fighter in the ring with both hands tied behind his back.

We knew that the punch (that is, the quota) was coming, yet were powerless, without the cooperation of NMFS to report the correct catches and claim a stake in the future allocations so vital to the recreational and, I might add, commercial fishermen of this country. Repeatedly, NMFS was asked to obtain yellowfin tuna statistics from coastal states. Sometimes a state would publish its own statistics, as did North Carolina, New Jersey, and Georgia. From 1995 to 2000, a picture began to emerge that the U.S. recreational yellowfin tuna catch alone was in the neighborhood of 18,000 to 22,000 metric tons or 39.6 million to 48.4 million pounds per annum. This was not the lowly 3 to 6 percent that NMFS would have the ICCAT managers believe, but represented a substantial fishery here in the United States EEZ, and it would represent from 11 to 13 percent of the ICCAT reporting just for the United States recreational fishing industry alone. Add another three percent for the commercial sector and one would have rights to 14 to 16 percent of any ICCAT quota passed in the future. All of us in BAYS committee meetings at first pleaded, then cajoled, then begged, and then demanded that NMFS conduct an accurate survey. Each year for the past five years we were stonewalled with such statements as:

> We are a minor player. There is no need to go further at ICCAT.

> We don't have the money for such a survey. We don't have the time.

> We did a survey this year, but there are holes because some states did not answer our letters about reported yellowfin tuna catches (six were missing). Therefore the report cannot be completed.

Excuse after excuse for not doing what NMFS is mandated to do under the fishery law of the land, the Magnuson-Stevens Sustainable Fisheries Act.

Then came Dr. Gary Matlock's famous statement when I confronted him with the minutes of the BAYS working group where I had listed, for the record, how many years the request for the yellowfin tuna data was part of the official documents. His voice rising in anger and pique, he said, "I am just not going to do it and that is that." He announced it at a full meeting of the ICCAT advisory panel in the fall of 1998. I gave him the nickname, "Dr. No."

All of us working in the recreational sector felt uneasiness in the pit of our stomachs. We knew that one day we would be screwed out of our rightful share of any ICCAT quota, but we did not know why; nor did we know who was behind this. The political pressure was coming from somewhere. As Satchel Paige once said, "Don't look over your shoulder too long or the door will slam you in the face." Little did we know what door. We all gathered in one of the minor conference rooms at NMFS, drinking coffee to try to fight off the tiredness from the 4 A.M. trips we just made to make this meeting. Precisely at nine Matlock appeared, strode in, walked to the front of the conference table and announced that something had come up and he had to go downtown to a meeting. "Sorry, but that is the way it is in Washington. I am going to send in someone to 'take care of you' in a few minutes. Good-bye." He turned and walked out. We were flabbergasted. Politics had struck again. The door was beginning to close, very quickly. In hindsight, I knew that someone was directing this scenario. No one would have crossed

Terry Garcia from NOAA without some powerful friends behind him. Matlock and many of the NMFS personnel were taking orders from someone. Normally a federal employee in a responsible job responds to political pressure. If you can arrange a meeting with a congressman who serves on the Senate Finance Committee or House Appropriations Committee, where the agency's annual budget is dissected and approved, you have hit the political hot button. Who would have enough clout to have NMFS stonewall reports, refuse to conduct surveys, massage stock assessments, and, when pressured to face reality, tell you to get lost. This is real clout, because no one would dare play such a dangerous game unless his backside was completely covered. Could there be such a person or entity that would have that much clout?

Soon it became all too apparent. Who has the most to lose and who commercially sold the most yellowfin tuna? The answer was in six hours of research that I did at the *Wall Street Journal* library in New York. The answer lay buried in a corporate report of a Fortune 500 company, and it will reveal itself to the reader in Chapter 13.

CHAPTER 11

Astillero Barreras
Vigo, Spain
175-foot Purse Seiner
November 1998
0900 hours

• •

I was standing in the expansive wheelhouse of a 175-foot Spanish purse seine fishing vessel docked at the Barreras shipyard in Vigo, one of Spain's largest commercial ports both for shipbuilding and commercial fishing activity, including a tuna cannery (Exhibit H at pages 205–210). Spain is the largest builder, in tonnage, of commercial fishing vessels in the world. I had just taken a tour of the ship, a masterpiece of mechanical marvels. I looked down into six vats thirty feet deep. They gleamed in the sun, polished spotlessly clean. Here, when the vessel was fishing, the catch was stored in a slurry of briny mush. If I had fallen into one of those caverns, I am sure I would have been killed. The gear, all hydraulic, left little to manual

74

labor in pulling in the purse seine net. Everything was ship-shape and in excellent condition. No rust bucket, this vessel. Before me, on the wheelhouse console, was a dazzling display of electronics, which the captain was explaining to fellow IGFA trustee José Luis Bestigui and me.

"Here is a scope where I can see the entire school of tunas (yellowfins or bigeyes). If I miss any, I can come back and complete the capture of the whole school." The vessel had a capacity of 1,900 metric tons, 4,180,000 pounds. I looked around the floor of the wheelhouse and saw about twenty-five cages, each the size of a soccer ball. "What are those?" I asked.

"They are GPS (global positioning system) devices that we attach, one each, to a fish attracting device or FAD. These FADs are made from bamboo or other such woods; are made into a pyramid shape, thirty feet high; and are covered with cheesecloth. They are designed to float in the current about sixty to ninety feet under the surface. In about a week, small critters get attracted to them. The ocean's rich plankton attach themselves to the wood and the cheesecloth surface of the FAD.

"In another week, small baitfish are swarming around the FAD, feeding on the smaller critters. In two weeks, small (two- to five-pound) tunas become attracted. If left alone for thirty to forty-five days, larger and larger tunas will feed around the FAD, and then it is only a matter of time before the apex predators, the marlins and sharks, show up to begin their feasting. Nowadays the latter never happens. As soon as the tunas reach about five pounds, the purse seine vessel takes a set with its net around the FAD and hauls in the net's contents, FAD and all. The GPS devices, one to a FAD, give the purse seine vessels accurate [to within fifteen feet] at sea positioning of every FAD that has been put out. The vessel can make larger catches because it can monitor the larger and larger ocean area."

"Where do you fish?" I asked.

"The Gulf of Guinea, off Western Africa," he replied, "off Sierra Leone, Nigeria, Cameroon, Liberia, and Ghana."

"How many boats in the fleet?" José Luis asked.

"A little more than fifty," the captain replied.

"How long do you stay out?" I questioned.

"Until we fill up the holding tanks," he said. "We make about four or five trips a year. One half of the crew is from the native country we fish off. We bring the fish to the cannery here at Vigo, although there is one in Ghana."

I quickly did the math in my head: 4,180,000 pounds times fifty vessels, times 4.5 trips. My God, that is 940.5 million pounds (427,500 metric tons) of tunas per year just from this area. If the average size fish was under five pounds, this would mean that 188.1 million small tunas were being killed each year. I quickly thought, as staggering as that figure was, there was worse news because these calculations did not include the bait boat catches. A bait boat throws live bait overboard, gets the school of small tunas in a frenzy and with a long bamboo pole baited with live bait or a feather, hooks a fish from the frenzy and heaves it over the fisherman's shoulder onto the deck where deckhands throw the still-thrashing fish into the hold of the boat, which may or may not be refrigerated. I have seen, at the port city of Faial in the Azores, bait boats (see Chapter One) come back to the dock with two inches of freeboard from the heavy load of skipjacks (a small tuna) aboard. The catch of skipjacks piled up on the deck, into every crevice and corner of the boat. As the vessel was unloaded, the bottom 30 percent of the catch was absolute mush and unfit for the market. The fish were crushed into a gurry and eventually sold for pennies to the farmers for fertilizer.

The ICCAT reports of yellowfin and bigeye tuna catches were far below the figures I did so quickly in my head. The entire ICCAT reported that yellowfin and bigeye tuna catches were 135,000 and 98,000 metric tons respectively for all areas, not just the Gulf of Guinea. It seemed to me that instead of going to the SCRS conference room and divining the statistics, the scientists should come here to Vigo and sample the seiners as they returned to the cannery in Vigo and also in Ghana. Two thoughts crossed my mind almost simultaneously: How can the oceans absorb this massive kill every year, and how shortsighted our own American fishery managers, directed by Dr. Gary Matlock, were in keeping an artificially low figure on our own yellowfin catches. I did the math as I had done many times during our own ICCAT advisory meetings. We had a United States fleet of 250,000 charter and recreational boats fishing for yellowfin tuna from Maine to Texas. If these boats averaged six yellowfin tuna a year per vessel at an average size of thirty pounds per fish, we would catch 45 million pounds (over 20,000 metric tons of yellowfins a year). This would represent about 12 percent of the ICCAT reported catch for the Atlantic and the Caribbean. I kept thinking of the meeting Matlock had walked out of the previous April. I knew that when the time came, American recreational and commercial interests would be sacrificed to some unseen god presiding over the Gulf of Guinea in Africa. But God sometimes reveals himself in vast and mysterious ways.

CHAPTER 12

Microwave Telemetry Headquarters
Columbia, Maryland
Rich Ruais, Molly Lutcavage
Cookie Murray, Steve Sloan
April 1999

· ·

I received a call from Rich Ruais, head of the East Coast Tuna Association, asking me if I knew anyone who could fund a school bluefin tuna tagging program. Rich is about six feet tall, wears glasses, and has a perennial, wide-eyed stare as if to say, "What's new?" I am sure he got this look from being constantly surprised at the machinations of the fishery business. He weighs anywhere from 210 to 240 pounds depending on his diet of the moment. The one thing that positively turns him on is the marbleized fat between the muscles of a giant bluefin tuna.

"Pure gold!" I have heard him say over and over again if he gets the right order from the numerous seafood restaurants we have dined in over the years of the ICCAT meetings. Rich is highly intelligent, a fierce advocate for U.S. fishermen's rights against a sea of troubles from the European Community (EC, later EU). Rich and I had started out to be bitter enemies, for, after all, he was on the other side of the Montauk tuna case described in Chapter One. After years of working at ICCAT together, we had begun to have a degree of mutual respect. I found him to be an authority on world fisheries, willing to discuss them with all of their warts and blemishes, and he found me willing to listen to reasonable arguments providing they contained no ideological agenda that was biased or unreasonable. I knew how much he was agitated (including one heart attack at forty years of age) about the mass killing of bluefin tuna in the Mediterranean by the fleets of the EC. He was particularly galled by the massive catches there of small tuna under five pounds in weight. The chance of a recent angina attack arose when in 2001 the EC refused at first to report its catch and then did so only at the very last minute, the day before the delegates to ICCAT were headed home.

Attending this meeting were Molly Lutcavage from the New England Aquarium in Boston and Cookie Murray, captain of the charter sportfishing vessel *Cookie II*, and me. We were meeting with Paul Howey, head of Microwave Telemetry, Inc. This company had developed an electronic pop-up tag for use on bluefin tunas. After being inserted in the fish, the tag could be programmed to pop up in the ocean sixty, ninety, or a hundred and twenty days later. The tag would then begin transmitting data to a satellite that would, in turn, transmit to a computer monitored by Molly. Rich, Molly, and Cookie had already tagged thirty-six tunas and had received a high incidence of return. We were meeting that day to talk about developing a smaller, miniature pop-up tag for small bluefins. The problem with using the existing tag was its size. The tag was about eight inches long with a bulbous end that would swing above and wide of the fish's skin at every undulating movement. The tag end never touched the skin of the giant tunas that weighed over five hundred pounds. If one used the same tag on a smaller tuna it would keep hitting the skin and rub the skin raw and kill the fish through bleeding or to sharks. See Exhibit I at pages 211–217 for the SCRS report on those tags.

The whole idea of being able to tag smaller fish was very important to the SCRS scientists compiling stock assessments of the Western Atlantic bluefin tuna. In fact, I had tagged one bluefin of about twenty pounds off Montauk one fall; it was recovered in the Bay of Biscay off Spain some three years later and weighed 125 pounds. Politically, it was extremely important for the American commercial bluefin tuna fishermen to disprove the existing two-stock theory under which ICCAT (and therefore all allocations) had been making judgments. The two-stock theory was devised by the European bloc some dozen years ago to keep the Canadians and Americans out of fishery management issues in the Mediterranean. The scientists, bowing to the political pressure from Spain, France, and Portugal, created this two-stock theory and this meant two-stock assessments. Allocations were given to the United States, Canada, and Japan for the Western Atlantic and the Caribbean stock. The Eastern Atlantic and Mediterranean stock allocations were given to the ICCAT European and Mediterranean contracting parties.

Turkey, Greece, Albania, Yugoslavia, Tunisia, and Italy, for example, do not belong to ICCAT and therefore do not officially report their catches. Even among the ICCAT European and North African member countries the reporting was sporadic. The scientific report showed that the Western stock of bluefins was in trouble from 1987 on, while the Eastern Atlantic stocks seemed healthy and therefore no quota was imposed. The United States and Canada reluctantly began to implement laws that fitted the science. ICCAT imposed a 2,800 metric ton quota in 1987-88 for the Western Atlantic, including our EEZ. The United States and Canada were the prime suppliers of quality bluefin tuna to the markets in Japan. The Mediterranean catch was not a material factor in those years.

By 1999, just a dozen years later, that catch in the Mediterranean had soared to 38,000 metric tons or nearly fourteen times that of the United States and Canada. I have a report from an impeccable source that, in addition to the known catch of 38,000 metric tons, for all the Eastern Atlantic, Libya has a private deal with Japan to supply an additional 75,000 metric tons of bluefin tunas. The drumbeats from the Mediterranean waterfront were sending a message that fishing is good here and the catch figures of 50,000 to 60,000 metric tons per year were being bandied about. Japan was everywhere, buying fish in the United States at seven or eight dollars a pound and at three to five dollars a pound for Mediterranean fish. Our Western Atlantic fish seemed to have larger fat content and therefore was more desirable in the Japanese marketplace. The Western Atlantic quota was being negotiated downward every year. The SCRS science reports showed that these bluefin tuna stocks had declined 90 percent from their 1973 historic highs.

The United States and Canada split the allocation roughly even-steven at about 40 percent each, and Japan picked up about half that or 20 percent more or less The scientists claimed that, while there was "some mixing of the two stocks, there was not enough to declassify the stocks into a one-stock theory." American and Canadian fishermen began to notice a slow decline in the size of large fish. In the 1980s the average size of the Western Atlantic stock was close to five hundred pounds. By 1999, this had dwindled to just under four hundred pounds, a sure sign the stocks are being overfished. In 2001, great pressure was being put on NMFS to allow the sale, for the first time, of bluefin tuna that weigh between 200 and 250 pounds. A bluefin tuna is not sexually mature until it is seven years old and weighs over four hundred pounds. These sexually immature fish fell into the category that were not for sale prior to 2001.

The United States and Canada passed strict rules to conserve these bluefin tunas, some of which were that every vessel fishing for bluefin tuna had to have a permit in the category it wished to fish (harpoon, purse seine, private angling, and charter boats) from which no fish could be sold, and a general category by rod and reel that allowed selling with a tag attached to each fish. All vessels fishing in any category must pre-register and obtain a tuna permit that went with the vessel. A landing tag that was pre-registered was put on every tuna. Japan agreed to cooperate by not buying any fish that was not properly tagged. Fish dealers were held accountable for each tag. The legal catch rate was one giant tuna per day per registered vessel. No offloading was permitted. This was a practice of catching one fish and passing it off to another vessel and keeping the next one in the boat. Both vessels had a tuna permit so it looked as if two or more boats had caught these fish when in reality only one boat caught the fish, and that made the second one illegal. In practice, an accommodation would be made between the owners of the vessels and the money from the sale was then split.

The fishing blooper of all time came from the press. A cover of the *National Fisherman,* the commercial fishing industry's monthly magazine, had a picture of one vessel backed up to another with a tuna hanging from the A-frame in the stern of the former ready to offload a giant tuna onto the other vessel. This, of course, was in violation of the law. Much giggling and razzing followed but no action from the enforcement arm of NMFS.

Molly Lutcavage from the aquarium in Boston, and Barbara Block from the Monterey Bay Aquarium were the deans of tagging bluefin tunas with the electronic tags; thousands from 1985 to the present. At first, these tags were not electronic and when recovered would only show data numbering days at sea, the estimated weight at the time of tag and of capture, and the distance traveled in a straight line. Thanks to Molly and some inventive scientists, new kinds of electronic tags were invented. The first has been described and the second, called archival tags, carried a wealth of information. These tags, with foot-long whip antennae, were inserted under the skin of the tuna and the fish had to be caught for the data stored in the tag to be used. But this stored data was very important and gave scientists a clearer picture of a bluefin tuna's world: Depths during the travel, metabolisms, speeds and total distance traveled, and a day-by-day, hour-by-hour location of the fish became available to scientists.

Very few nonelectronic-tagged fish were reported from the Mediterranean, probably because of the suspicions of the European fishermen that this science would somehow come back to haunt them. The pop-up electronic tags needed no recovery system from fishermen. Once the pop-up electronic tag pierced the surface of the ocean, the data stored inside began transmitting to the satellite overhead and then to the computers that Molly was monitoring in Boston. Almost immediately, several tags popped up off Sicily and the Bay of Biscay near Spain. This got Rich Ruais and others thinking about disproving the one-stock theory. If he were right, all of the conservation efforts, the quotas, the political forces in play, for naught. The American fishermen had obeyed the law, followed the quotas and then saw their fish being taken in a ratio about ten to one by Mediterranean fishermen who operated without quotas, and without tags and seemed to have an unregulated reporting system. Even worse, when the subject popped up (please excuse the pun) at every ICCAT meeting, the United States commissioners were told to go home, mind the Western Atlantic, and take care of their own tunas. "Don't tread on us," was the message. The pop-up tag was a foolproof way to make the case. In the meantime, slowly but surely, scientists from SCRS began to evaluate catch data from the Mediterranean and concluded that the massive killing there could not continue at the rate reported.

My radio show *The Fishing Zone* is carried by the Talkone Radio Network and is broadcast to two hundred stations and reaches one thousand cities and an unlimited number of people who can listen on the web by going to the website www.talkone.com if they have live audio. I broadcast on

Saturday morning at 6:05 Eastern time. I have been on the air for 583 (as of this writing) consecutive live broadcasts. I review fishery matters, ecology, conservation, scientists, and authors who write about these subjects. During October 2000, I reviewed the book *Mattanza*, by Theresa Maggio. It was the true story of a small island off northern Sicily called Favignana, where a two-thousand-year-old tuna trap was still taking giant bluefins. The book told the tale of the history of the men, the trap, and the fishery. At the end of the book, there was a roll call of some fifty traps that had become useless because of the lack of fish. I asked Theresa to call the roll in her mellifluous, lyrical, melodious Italian. She did; it was a memorable and magical Saturday morning, tinged with sadness. I felt that an era was gone, never to be recovered, especially since I have observed the EC delegates from ICCAT connive their way around implementing regulations designed to save these magnificent fish for future generations.

As of this writing we are still working on making the smaller tags for the juvenile bluefin tunas. The kill is still out of control in the Mediterranean, and a new threat to these magnificent fish called IUU fishing (illegal, unregulated, unreported) is taking place. Everyone wants the money. Japan is buying, and the world is selling against its own self-interest.

At the 2001 ICCAT meeting in Spain, the EC refused to submit its catch data until the last hour of the last day of seven days of meetings. It admitted to taking 34,000 metric tons, 6,000 metric tons over the limit. The United States refused to adopt that report and insisted that the EC come into compliance with a reduction of 6,000 metric tons for the 2002 season.

The EC refused to accept this. The meeting broke up and all delegates headed for the airport with the EC out of compliance. I would love to report that an agreement has been reached. It has not.

CHAPTER 13

Office of Stephen Sloan
230 Park Avenue, New York
April 1999
1900 hours

• •

My computer was online. I had just received my e-mails and noticed that the Marine Fish Conservation Network had posted a notice reporting that the H. John Heinz III Center for Science, Economics, and the Environment was forming a committee to study the reauthorization of the Magnuson-Stevens Act. The Center was named for the United States senator from Pennsylvania who was killed several years ago in a plane crash, The committee was to consist of Dr. Gary Matlock and Dr. Andrew Rosenberg, both from NMFS; Dr. Susan Hanna, Heather Blough, Richard Allen, Suzanne Iudicello, and Bonnie McCay. Their qualifications are listed here alphabetically:

Richard Allen was a commercial fisherman for thirty-three years, based in Point Judith, Rhode Island. He was a commissioner of the Atlantic States Fisheries Commission for ten years. He served on the Sea Grant review panel.

Heather Blough received a degree in marine biology from the University of North Carolina and a graduate degree in environmental science and policy at Johns Hopkins University. She is a researcher and writer on marine projects at the National Academy of Sciences.

Dr. Susan Hanna is a professor of marine economics at Oregon State University. She was on leave to the Heinz Center during 1998-99 to direct the Managing U.S. Marine Fisheries program. She has served on many boards including several within NOAA, and also is on the committee to review individual transferable fishing quotas. She is the author of numerous articles and three books about natural resources and their management.

Suzanne Iudicello is a researcher and writer and is the principal of Junkyard Dogfish consulting firm. She serves on MAFAC (the Marine Fishery Advisory Committee chartered by Congress and run through NMFS; I served as its chairman for three years). She was a specialist for the Center for Marine Conservation, a special interest group in Washington, and a member of the Green Team on conservation. She has written widely and she represented Alaska in the Washington office of its governor.

Dr. Gary Matlock is director of Sustainable Fisheries for NMFS, the number two job in NMFS. Before NMFS he was with the Texas Park and Wildlife Department, and has published widely on fishery matters, biology, and sociology.

Bonnie J. McCay is a professor of anthropology in the Department of Human Ecology at Rutgers State University. She served on many advisory committees and has published many articles on the culture of the commons, participatory management, and property rights of citizens.

Dr. Andrew Rosenberg, while not listed in later committee documents, served in the initial formation. He was regional director of the NMFS New England office and left that post in 1999 to go to NMFS headquarters in Silver Spring, Maryland. From there, in short order, he took up a post at NOAA and shortly thereafter left to go to the University of New Hampshire's marine science grant programs.

It is apparent that no one represented the rights of recreational fishermen, the largest user of marine resources, on this panel formed to study the Magnuson-Stevens Act, which formed the basis of the United States marine public fishery resources.

The reauthorization of the Magnuson-Stevens Act was of great importance because all United States fishery laws and regulations, plus the management practice and, most important of all, the funding stemmed from this act. I wrote a letter on June 28, 1999, to John Sawhill, chairman of The Heinz Center, suggesting that recreational fishermen be represented and also inviting him to appear on my radio show, as shown in Exhibit J on pages 218–219. On July 12, 1999, I received a terse letter from Dr. Hanna, program manager, saying that recreational fishermen had been consulted, that a report would appear on the Heinz Center's website, and deferring the radio show for the time being.

Having just come from the disastrous NMFS April meeting with Dr. Matlock, I got an uneasy feeling in my gut about this newly formed Heinz committee, its membership, and its role in reviewing our most important fishery laws. I became more uneasy when Matlock, in October, just before announcement of this formation of the Heinz committee appeared on the Internet, refused even to consider reviewing the amount of yellowfin tuna the recreational sector caught in 1997-98. I asked myself the ultimate question: What were the third- and fourth-highest officials of NMFS, whose salaries were being paid by the American taxpayers, doing serving on a not-for-profit commit-tee/panel whose apparent mission was to supply information and publish a position statement about amendments to the law (the Magnuson-Stevens Act) directly affecting the government's funding of hundreds of millions of dollars for the NMFS annual budget? I became increasingly uneasy the more I thought about the Heinz-dominated committee and its objective, knowing as I did that fisheries were dominated by commercial interests at NMFS. Hanna's letter and Matlock's belligerent actions gave me no comfort that recreational fishermen would be treated fairly in this committee's report. That Teresa Heinz, John Heinz's widow, ran the foundation that received millions from Heinz 57, a public corporation in which she was a major stockholder, and that she was married to Senator John Kerry of Massachusetts, also gave me considerable angst. Senator Kerry never agreed to a bill or order that cut back on any kind of commercial killing, and for whom the phrase "recreational fisherman" meant some-one from the outer reaches of Tibet.

I decided to play my hunch to the hilt after Matlock's October massacre of recreational yellowfin data, and upon my return from ICCAT and the visits to the Vigo shipyard, I became convinced that someone was directing NMFS to act in such a highhanded way toward recreational fishermen. A seat-of-the-pants, quick study done of the American Sportfishing Association's report done in 1996 showed that the true economic value of recreational fishing amounted to an output of fifty-two billion dollars per year (Exhibit K at pages 221–232). That calculation did not consider the ripple effect on hotels, restaurants and other food vendors, and other ancillary businesses as well as property taxes paid on marinas, boats, and waterfront businesses. Clearly this proved that a fifty-fifty parity should exist with U.S. commercial fish landings. This parity should extend to grants, full-time jobs, and agenda items.

One problem was that NMFS never quoted anything in an official document that equates recreational fishing in dollars, only in effort. How many fishermen, how many trips, how many days spent fishing? The commercial industry was always quoted in dollars for its catches and landings as well as imports. The message was clear and effective. If NMFS could keep the two in twain, most of the monies, grants, and council seats could be stacked on the side of the dollar producer, that is, the commercial sector. The NMFS strategy is that if we can keep recreational fishermen quoted in effort with no dollar signs attached, commercial fishing will dominate all sectors. After all, I mused, NMFS was the Bureau of Commercial Fisheries until President Nixon changed the name in 1973 to the National Marine Fisheries Service. He may have changed the name, but he could never change the mindset of the agency.

I followed my hunch and in February of 2000 I sent in a Freedom of Information Act (FOIA) letter asking for a five-year history of any grant requests made by the Heinz Foundation and a list of grants that were funded. (See Exhibit L at page 233–241). Bingo! The Heinz Foundation got a $250,000 grant for the Committee to Study the Reauthorization of the Magnuson-Stevens Act, also called the Sustainable Fisheries Act. According to my theory, Matlock and Rosenberg double-dipped into the taxpayer's porridge bowl. They were paid by NMFS to create such a report, then as a political favor, put through a grant for a quarter of a million dollars to produce a report that the service should have done in the first place. The issued report would be a reflection of what NMFS wanted in the reauthorization of the Act anyway. Why go through this elaborate screen, using Teresa Heinz to influence the pending legislation on reauthorization? Why did the Heinz Foundation refuse to place anyone on the panel who had deep knowledge of the recreational sector, which was an apparent coequal in the pending legislation? The Heinz Foundation was funded by monies from Heinz 57 whose balance sheet shows billions in capitalized value. Why did it have to tap NMFS for $250,000 when it could have funded this by itself?

One day I went to the *Wall Street Journal* library here in New York on 34th Street and looked up Heinz 57 Corporation. To my astonishment, I found in its annual report that Heinz 57 sold over one billion dollars a year in tunas and tuna products, most of them being yellowfins. Heinz 57 also caught and processed large amounts of marlin in the Pacific for cat food for one of its brands; see Exhibit M at page 242. In mid-September 2000 and continuing into the June 2001 quarterly earnings report, Heinz 57 reported a 3.5 percent decline in earnings because of the lowest tuna prices in thirty-four years. There was a glut of small tunas on the market; a small wonder with all of the purse seiners using FADs in the Gulf of Guinea. Heinz owned StarKist tuna products as well as many other fishing divisions. This was a case of a public company having direct input into a governmental agency using a foundation as a screen. There are other brands of tuna and catsup to buy in case you agree with me. At the very least, the Heinz Center should refund the grant money it received, with interest, having committed an egregious conflict of interest with NMFS and Heinz 57, adding new dimensions to the term. Dr. Matlock should be brought to task for his part in the taxpayer-bilking scheme. He is now buried in the bowels of NOAA, after giving up his position as head of Sustainable Fisheries at NMFS. Dr. Rosenberg, for his part in this misappropriation of the public trust for a period of fifteen years, should be prohibited from soliciting grants for the University of New Hampshire, where he is presently ensconced. There is more. The ICCAT delegate from Ghana handed me his card at the 2000 Marrakech meeting. He now represents Heinz 57 and its canneries in Ghana, which is the closest port to the Gulf of Guinea, where the massive catch of yellowfin and bigeye tunas takes place. I assume by now that you are not surprised one whit at this information

In June 2002, Heinz 57 announced that it was selling or merging the StarKist division to Del Monte Foods in a billion-dollar deal. I wonder if this sudden divestiture is not part of a plan to back Senator John Kerry's bid for the presidency in 2004?

CHAPTER 14

United States Delegation Room
Othon Palace Hotel
Rio de Janeiro
November 1999
1900 hours

• •

"When sorrows come, they come not single spies, but in bat-talions," said Hamlet. Bad news kept piling up against the United States's ICCAT negotiating positions. One by one, our delegates reported to Rollie Schmitten, our government com-missioner and head of the American delegation, that fixed ele-ments of our 1997 billfish agreement now were not fixed. We were relying on 1997 Document 71, Exhibit F, which was redrafted at the last minute. It now seemed, and the Japanese claimed, that we had to agree to further cuts of recreational bill-fishing in the United States, which was contrary to the United States's position. If this happened in private industry, somebody's head would be on the block. Even worse, Japan was now insisting that the United States reduce recreational effort (days at sea) as well as landings (fish on the dock). This was quite disingenuous because 95 percent of the recreational billfishing had been on a voluntary hook-and-release basis. In fact, for the reporting year of 1998 the recreational community landed only just over two hundred billfish for all of the Americas including the Caribbean. On further inspection, this was the average for the previous ten years. This was payback in the form of stick-it-to-'em time. Japan was getting even for our having the audacity to insist that billfish mortality be cut on longline vessels.

The huge American boating industry is unique. Nowhere else in the world does the ordinary working person have the chance to own a boat as in the United States, and many boats are bought just for the opportunity and enjoyment of being out on the ocean looking for billfish. But recreational boating and fishing participants are looked upon as "fat cats" by many bureaucrats and staffers at NMFS. These recreational partici-pants are regarded as standing in the way of pure fishery man-agement and the business of catching, killing, and selling and not releasing our living marine resources.

During 1997, SCRS studies reported that white and blue mar-lin were overexploited and had fallen severely below maximum sustainable catch, possibly affecting their ability to have any stock recovery at all. Nelson Beideman, blue water industry longline spokesman, reported to our delegation that there was no movement in getting the major players to reduce their land-ings of swordfish from 13,000 metric tons per annum to between 10,000 and 10,600 metric tons. Superimposed on this was an active Green Team environmentalist NGO (nongovernmental organization) in attendance, whose spokesman was beating the drum for 9,000 metric tons maximum in order to ensure that the intended recovery would actually be achieved. When Nelson was giving his report, he stated he could give the Japanese some of his unused quota of southern swordfish. I kept wondering why recreational fishermen had to rely on commercial intermediaries to take up the billfish conservation cause. I knew that Nelson's best and sometimes only customer was the minions from Japan who bought his fish. Why wasn't our own government fighting diligently for the protection of the huge recreational boating and other businesses in our country? After all, wasn't NMFS a part of the Department of Commerce, the supposed protector of American business interests worldwide? This question is still out there today, begging to be answered.

Another report came from Kim Blankenbeker, our capable delegation executive secretary, that stated that the European Community refused to make data available for small bluefin tuna catches in the Mediterranean, even though we were into the third day of the meetings. This was a clear violation of the agreements. To top it all, six of our ICCAT delegates were mugged right outside the hotel, and the State Department issued a warning in a meeting as follows: "If you see a band of children approaching you, give them all your money without any questions asked. Do not provoke anyone." It seemed to me that not all the muggings were taking place on the street. As postscript to the week, several shots were fired in a nearby beachfront café on the last evening of our stay. Several of our delegates in attendance hit the floor.

Yet, amid all this confusion, a sea change was occurring. I noticed that Japan's delegates were being unusually solicitous toward members of the European Community delegation in the open plenary sessions for the first time in six years. Seemingly this was because Japan does about ten times the business in bluefin tuna, swordfish, and bigeye tuna with the EC than she does with the United States. To put a topper on all of this intrigue, Brazil, the host country, was making loud statements about its sovereign right to extend its EEZ along its coastline. Joint ventures were being made and talked about at the meetings. Brazil wanted more allocation from the present bluefin tuna and swordfish quotas. No one had the answer to the question: "Where does the Brazilian [and other developing countries'] quota come from?" A pervasive and emphatic answer from everyone with a quota was: "Not from my quota." All in attendance agreed to try for a solution at a later date. This came in the form of a second Intersessional in Brussels in May 2001 dealing with the "criteria of allocations."

There were also compliance issues to be addressed by the delegates as reports came in about IUU fishing and foreign flag-of-convenience landings and catch allocations. It has been an established fact that, for many years, emerging economies of poorer countries made some hard currency by allowing vessels to be registered in their countries, to carry their flags, and to pay a fee for that privilege. The advantages were many: lower insurance standards and lower inspection standards, few if any union contracts and therefore lower crew salaries, obfuscation of landing and catch data, making it harder to inspect the vessels and account for the landings. In all, it is a more cost-effective operation; cheaper, but more dangerous for the health of the worldwide fishery. Foreign-flag vessels, I learned from one newspaper investigative reporter, apparently carry 75 percent of the illegal international arms and drugs trades. The following is a typical happening.

Vessel A is owned by a Spaniard; sails with a Spanish captain. The vessel is fifteen or more years old. Instead of retrofitting the vessel to meet insurance standards, the vessel is reflagged in Panama, for instance. That reflagged vessel can now hire low-paid Bangladeshi seamen. The vessel catches fifty metric tons of Atlantic bluefin tuna and pulls into Tobago to unload. Under whose ICCAT-granted quota are these fish caught? Where does this catch get reported in the ICCAT system? Is it the national register flag of Panama that the vessel is flying at that moment, or the country where the vessel was built, or the domicile of the owners, or the country that has the quota granted from ICCAT, or none of the above? The answer is Panama or none of the above. Meanwhile, the Spanish owners, rubbing their hands with glee, go out and build a new boat to catch Spain's share of the quota once again. Panama, which has no quota, becomes an IUU country, but no letters ever go out and demand a stop to these proceedings. How can the scientists do reliable stock assessments if this happens on a large scale and regular basis?

At the May 2001 Intersessional in Brussels, I later submitted a memo to the secretariat and the delegations of Japan, the United States, and South Africa, on how to solve the foreign-flag problem on fishing vessels. This will be discussed in the chapter on solutions. Inside our own delegation at Rio, a snickering campaign began whenever Jimmy Donofrio, from the Recreational Fishing Alliance (RFA) asked to speak. Some delegates were rude and extremely disrespectful. These government employees involved apparently felt, "Hey! This is our show. We have jobs to protect and we are not going to let anyone else type a page or prepare papers. We are in charge here and no one is going to take away any authority or turf from us." It got so bad that Jim and I went to their departmental superiors and complained, and this behavior slowed down, but never stopped. It was also very counterproductive. Private industry delegates do not get paid for their time by our government, only for their hotel and travel. Nine long days are donated by private industry user groups as well as by the environmentalists for each ICCAT meeting plus six to eight days in the United States before and after. Somehow, certain members of the NMFS staff paid little or no respect for this contribution. I know I felt that I was responding to President Kennedy's call, "Ask not what your country can do for you—ask what you can do for your country."

During the first four days of the ICCAT meetings, all of us on the delegation, private and public, felt that this might be the end of any serious fishery conservation work. Stalemates existed on all fishery issues, and many of us felt that perhaps ICCAT has outlived its usefulness and has ceased to function effectively in the way that had been planned.

CHAPTER 15

United States Delegation Room
Othon Palace Hotel
Rio de Janeiro
November 1999
1800 hours

• •

It was 1999, the year of the swordfish at ICCAT, and this annual meeting was held in Brazil, the first departure in five years from meeting as usual in Madrid. These meetings are not some Congressional spend-the-public's-money junkets. They are hard work, starting each day at a seven-thirty breakfast that is always the same fare, talking fish all day with delegates, and ending most nights at midnight drafting responses, plenary floor recommendations, and memos. ICCAT is a World Fishing Congress for those thirty countries that have signed the North Atlantic Tuna Treaty and are called contracting parties. Each year these countries concentrate on one of the Atlantic and

Caribbean's highly migratory species (HMS) that make up the scientific monitoring list. The meetings are occupied with specific discussions by panels. Panel One being swordfish (SWO); Panel Two, bluefin tunas (BFT); Panel Three, such species as bigeye tuna, albacore, yellowfin tuna, and skipjacks (BAYS), and Panel Four, other billfishes and sharks. There are other panels and working groups such as Compliance and Permanent Working Group (PWG), the former dealing with the reports of the signatory nations and the latter with membership, fishing quotas, and general financial health (most times ill-health because eleven countries owed over $3 million in back dues in the year 2000, for example). And finally, the Standing Committee on Research and Statistics (SCRS), manned by scientists from many of the contracting parties. Each year a member is required to report his landings in the HMS categories.

In dollar value both the bluefin and the bigeye tuna fisheries are the money catch. The former represents some 40,000 metric tons or 88 million pounds, and the latter 93,000 metric tons or 204.6 million pounds in annual reported catches. The bluefins have a landed-on-the-dock market value of $7.00 per pound or $616 million while the bigeye tunas at $5.00 per pound have a value of over a billion dollars, $1,023,000,000. These numbers are before any markups occur as the products get distributed into the system. For an example, the belly of a bluefin tuna, sliced by a scalpel to a thin wafer and placed on a rice ball to be served at some chairman's party in one of the corporate suites in Tokyo is worth $300 per pound. The buyers in 95 percent of the cases are the fishmongers of Japan. It is entirely conceivable that the combined economic value for bluefin and bigeye tuna products in Japan alone approaches $20 billion. In January of 2001, one 400-pound bluefin tuna sold in Tokyo's Tsukiji market for $172,000 or $430 per pound. It was reported that the fish was caught in Japan's home waters in the Sea of Japan.

Each year scientists on the Standing Committee for Research and Statistics meet several months before the general meeting to study the fish landing reports supplied to them by the contracting parties at ICCAT. These scientists are obligated to advise the PWG on trends regarding fish stocks of the monitored ICCAT species. Many people believe that these scientists are responsible for extremely accurate data. This is not so. The data supplied by each country may not be totally accurate in and of itself, but the statistics are gathered from enough diverse sources so that trends will emerge for hopefully intelligent decisions regarding the stocks and catch allocations that should be made by the SCRS scientists for presentation to the full body at the ICCAT plenary sessions. The SCRS review updates the size of present catches with their previous estimates of stocks, the amount of fish in the Atlantic Ocean and Caribbean for each species of fish. The scientists are guided in their work by the principle of Maximum Sustainable Yield (MSY). This is the simple theory, for example, of having money in the bank and living off only the interest payments. It is the responsibility of every contracting party to submit catch data for each species controlled by ICCAT in that reporting year in order to be in compliance with the ICCAT's Contracting Party Agreement. But parties may be out of compliance as far as catch rates are concerned, as we will find out later in this story.

Looming over all of this process, like the Colossus at Rhodes, is illegal, unregulated, and unreported fishing (IUU).

This is done on a large scale by many countries such as Belize, Panama, Honduras, Trinidad and Tobago, Iceland, and Greece, and by the Faeroe Islands, and some vessels of the Taiwanese fleet. Many other contracting parties, including Japan, Spain, France, and Portugal, have regularly been accused of such IUU fishing. But the whole apparatus is a sham. Vessels fishing under a flag of convenience, such as the Panamanian or Honduran flags, are vessels previously built in Spain, France, or Taiwan. The money for the catch goes into the bank accounts of the owners who sit in their offices in Madrid or Marseilles or Taipei and watch their bank accounts grow as the funds are wired from the buyer (mostly Japanese) the minute the catch hits the dock and is loaded quickly into nearby trucks. The catch, under ICCAT allocation systems, is then charged to the flag flying from the vessel and not to the nation of the real owners of that vessel. As the bank account gains in value a second vessel is contracted for, and that newly built boat will now fish under an ICCAT member quota that is part of that country's allocations. This is double-dipping allocation maneuvering. The flag is not fishing; the owner is.

One could ask Canada why it fired on and seized a Spanish vessel that was fishing illegally in Canadian waters for turbot, a flounder-like protected fish. The Canadians found two sets of books and a false bottom where more illegal catch was stored on the Spanish vessel. The story was headline news in the Canadian press for weeks. Because of the clandestine nature of this kind of fishing, no one knows how large the IUU catch is. The scientists plug in a number that they estimate for each year's stock assessments, but in private they all feel that it is much too low. One might ask the question: If there is IUU fishing, where do they sell their products? It should be a simple exercise to issue a statement, "I will not buy from an IUU fishing fleet or vessel." "Follow the money," as Deep Throat from Watergate days said.

The money trail leads directly to Japan's Tsukiji market where millions of pounds of ICCAT parties' fish are auctioned daily. Unless ICCAT formally (this means a unanimous vote on the floor at the plenary session) declares a country guilty of being an IUU party, Japan feels free to purchase the seafood with impunity. The contracting parties at ICCAT fully understand that each one (except the United States and Canada) has a segment of IUU fishing within their fisheries. This declared policy of self-policing, or blowing the whistle on oneself, always fails, and purposely so, lest the spotlight of IUU fishing falls on one's own practices.

As I sat in the United States delegation room in Brazil, I began to wonder about the thick book of fishing statistics that I held in my hand. What if there were a massive amount of IUU fishing, even greater than estimated by the SCRS reports, and what if the contracting parties hedged their catch reports and kept the landings figures low? There was already plenty of evidence of that procedure, for, in fact, the European Community refused to report the total amount of small bluefin tuna catches in the Mediterranean. There was a report issued by the scientists that 50 percent of those bluefin tuna catches were under five pounds in weight. These small fish are called "zero-age fish." Not only were the parties taking massive amounts of spawning tuna, they were killing the future stock, the juveniles, as well. Rumors began circulating among the delegates that the

104

EC had caught 8,000 metric tons or 17.6 million pounds over their quota for 1998. In the year 2001, with its mad-cow disease, foot-and-mouth disease, and anthrax scare in Europe, there was a greater demand for protein to replace all the cattle that were destroyed. The one great source left was the oceans, already under severe pressure in so-called normal times. Can the world's oceans take such an onslaught year after year? I think not. The reports are always one year behind because it takes several months to analyze the data before the annual meeting. Under the 1999 agreement, the EC was to deduct 8,000 metric tons from the 1999 catch, going down from 35,000 to 27,000 metric tons. There were even headlines from the Department of Commerce report. In dollar value this meant that $123.2 million would be lost for EC for the year 2001. Clearly, it would be doing everything in its power not to sustain this loss of income. The feedback from the EC delegation was to be "damn the science, full catching speed ahead." In fact, the EC designed a way around this problem. It flat-out refused to submit the catch data; proving my equation correct: Science without compliance leads to defiance. As I said, 1999 was the year of the swordfish at ICCAT and SCRS scientists stated that for a 50 percent probability to return the swordfish stocks to maximum sustainable catch levels, the total quota had to be lowered to somewhere between 10,000 and 10,800 metric tons, down from 14,000 the year before. In pounds, this was equal to 22 million to 23.7 million, from 30.8 million respectively.

Swordfish are caught by longline methods. That is to say that a vessel lays out thirty to a hundred miles of line with a thousand to five thousand baited hooks each night. The average number of nights spent fishing by these longliners is a hundred and fifty annually. The total reported ICCAT fleet size is 1,460 vessels. Swordfish bring about $3.00 to $4.00 per pound at the dock. While swordfish are the primary target of the ICCAT longline fleet, the baited hooks, drifting in the current at night, catch many other species such as sharks, the marlins (white and blue, and sailfish), small tunas, dolphin fish, endangered turtles, and sea birds. These other species are called by-catch because they are not the targeted species of the swordfish longline vessels. In the last decade the average size of a swordfish landed by these longline vessels is under eighty pounds. These small fish have not spawned even once. Scientists tell us that catching so many juvenile fish is a sure sign of trouble among the spawning stocks of that species. These undersized swordfish gave rise to a very successful campaign by the conservation community to "Save the Baby Swordfish." Many seafood restaurants signed on and refused to serve swordfish.

The battle in Brazil went on for most of the week between the delegations of the United States, Canada, and the EC. The United States, Brazil, and Taiwan, and the EC represented chiefly by Spain, are the major swordfish players. Spain is included because the Spanish have built the largest fleet of fishing vessels in the world. On my visit to a shipyard in Vigo, Spain, I saw with my own eyes a 375-foot-long purse seiner being built. (See Exhibit H at pages 205–210 for illustrations of purse seiners and longline vessels).

The capacity of 3,000 metric tons was 50 percent larger than any purse seine ship afloat, equal to 6.6 million pounds. I asked the owner what the payback on this vessel was. He answered, "two years." I looked at him in disbelief.

"You mean to tell me that you can pay off the ship's cost of $26 million and all expenses in two years and then have a business free and clear of all debt?"

"Yes," he replied, "I am building a second vessel when this one is launched."

"I do not doubt it for an instant," I replied. Financially translated this means that after two years, the operation would make $13 million a year with no debt. The financial impact of these numbers got me thinking that perhaps there was another way to evaluate and support the present SCRS fish stock assessments that govern ICCAT's quota system, through the normal profit and loss reporting of commercial fishing activities. This method would eliminate the inherent conflicts of interest when the science is always in conflict with a nation's self-interest, especially in years of declining fish stocks, compounded by the clashing socioeconomic aims of the countries involved. As Sir Walter Scott said in *The Antiquary*, "It's no fish ye're buying, it's men's lives."

CHAPTER 16

United States Delegation Room
Othon Palace Hotel
Rio de Janeiro
November 1999
1900 hours

• •

The battle was still raging over the amount of swordfish the contracting parties could catch in ICCAT's Southern Atlantic and Caribbean zone so that there was a 50 percent chance for the recovery of the stocks. Everyone had given up on keeping the previous 13,000 metric tons (28.6 million pounds) quota of which the American longliners receive 29 percent. The contest had narrowed between accepting 10,000 or 10,800 metric tons (22 million to 23,760,000 pounds) in total as the quota for all contracting parties. While we debated over this seemingly narrow range of swordfish quota, news began to filter in from the European Community that a previous agreement for Eastern bluefin tuna was in error. The catch in the Mediterranean of bluefin tuna in 1997 was 35,000 metric tons (77 million pounds). The EC had agreed, as the SCRS scientists had recommended, to reduce that amount to 28,000 metric tons (61.6 million-pounds) per year.

The agreement went on to state that should the EC go over the 28,000 metric ton quota then that amount would be deducted in the next year from the base of 28,000 metric tons. As we sat in the delegation room, the EC's bluefin catch report, after much stalling on the fourth day (it had been due before the conference began) indicated that the EC had caught 33,000 metric tons; 5,000 metric tons (11 million pounds) over the limit. Fifty percent of that overage catch was mature fish, averaging four hundred pounds. This would equal 13,750 large tunas. The other half of the average catch would amount to 2,750,000 juvenile fish; average weight being two pounds. To put this figure in perspective, the overage of small fish was many times the amount of the bluefin tuna quota allowed for the United States from the Western Atlantic. By any stretch, this was a huge overage, and considering the methods of reporting and the reluctance of the countries to report such violations, these figures may be 50 percent or more too low.

Behind the scenes, Japan kept buying all the large bluefin tunas, but most of the small fish go to the canneries in Marseilles, France; and Vigo, Spain; and Ghana, in West Africa; and others I do not know about. The French, with good reason, are particularly touchy about anyone bringing up such mundane points as sticking to their agreements. Against this large amount of overfishing for bluefin tuna in the Mediterranean came the startling news that the agreement hammered out the year before, requiring the reduction in next year after overage was reported, was stated by the EC to be "flawed in language and concept." The EC stated, "The agreement did not take into consideration any underage for any years before 1997," and inferred it would go back to pre-1990. This clearly was never on the negotiating table. One must, begrudgingly, admire the cunning of the EC's lawyers. Rollie Schmitten, head of the U.S. delegation, was stunned. Rich Ruais, head of the East Coast Tuna Association, was in a state of shock because the U.S. commercial and recreational fishermen had made substantial conservation sacrifices for many years that had allowed the EC and its commercial fishermen to make such massive catches almost eleven times the U.S. position.

It was not until the last day of meetings and on the way to the airport, that the issue was seemingly resolved. The EC and the United States and the others agreed that the slate would be wiped clean, all accounts brought to zero. The next report due would be November 2000 at ICCAT in Marrakech, Morocco. During the year between meetings, November 1999 to November 2000, Rollie Schmitten confirmed this agreement three times in writing that all previous overfishing accounts would be reduced to zero. What happened here only reinforced my view that the United States delegation needed more legal help vetting our agreements. In the rush of the last day of meetings each year at ICCAT, things can slip because of the hurry, the translations, and the time pressure. It only took the first day of meeting at Marrakech in 2000 to confirm my worst fears.

CHAPTER 17

Offices of American Sportfishing Association, Washington
Present were Nelson Beideman, Glen Delaney, Robert Hayes, Mike Nussman, and Dr. Ellen Peel
Several days in June 2000

• •

The Longline Five were meeting in private. The group's membership consisted of the two private-industry ICCAT commissioners, a fishery lawyer, plus the head of the longline Blue Water Fisherman's Association and the head of The Billfish Foundation. The purpose of the meeting was to sign a secret Memorandum of Understanding (MOU) concerning a bill they wanted to introduce in Congress advocating the buyout of sixty-plus American longline vessels. The commercial commissioner was Glen Delaney, a registered lobbyist who served on the staff of Senator John Breaux of Louisiana and for a time was in the

import and export business of exotic animals. Glen enjoys fishing for fun, and Senator Breaux served on IGFA's advisory
board during the conceptual part of the development of the
Fishing Hall of Fame and Museum.

Glen is a powerfully built man now in his thirties. He is
always well groomed, has a first-class mind, and knows the
working of Congress and fishery laws cold. Many times he
helped recreational fishermen in the ICCAT meetings, but there
was always a certain reserve and guardedness about his work.
His main fault is that the more he is off base about some negotiating point, the more strident he becomes about getting it
through. I must grudgingly admit that if the recreational fishermen of this country had such an advocate in their corner things
would be a lot different at NMFS and among members of
Congress. I always have had the feeling, like England and
France dealing with the head of the Ottoman Empire during the
early 1900s, that there were those who wanted to keep the
recreational fisherman's representation as weak as possible. In
my opinion, Glen was very much part of that process during the
past seven years. Another member of the longline group was
acting vice president of the American Sportfishing Association
(ASA) Mike Nussman, who represented most of the fishing
tackle dealers in the United States. Mike is a former staffer for
South Carolina's Senator Fritz Hollings and is also the recreational commissioner at ICCAT. Mike's problem always has
been that he is pulled in too many directions at the same time.
This comes as a work habit from the staff's frenetic pace he set
for Senator Hollings. Compared to his predecessor, Mike
Montgomery, Nussman was a huge improvement; however, the
job had also expanded. Many times he had to be late for meet-

ings. It was particularly galling to other recreational members of the ICCAT Advisory Committee that our commissioner was not in attendance when the governmental and commercial ones were sitting at the head of the table discussing our problems without our own man there. I got along well with Nussman until this secret meeting. I told him I would never make an end run around him and would support him as long as I was kept in the loop. Nussman had an infuriating habit of talking to you and looking over your shoulder at the same time, as if waiting for some senior senator to appear from the cloakroom. His favorite phrase was, "I tell this to you in confidence, but I will deny it if asked." Through his training, everything seemed so secretive. This is the life of a staffer inside the Beltway.

The third member was Bob Hayes, a lawyer who, before becoming a partner in his own law firm, was a partner at Bogle & Gates whose clientele included commercial fishing interests. Prior to 1998, Bob Hayes came to many ICCAT and other meetings on behalf of his commercial fishing clients, and Bob Hayes was the one person whom recreational fishing interests really feared and mistrusted. For some reason, NMFS seemed to listen to him and then invariably side with his point of view. It was as if he had some pipeline into the catacombs of NMFS and, in fact, he had. His daughter was on the legal staff of NOAA, the NMFS parent.

Bob left Bogle & Gates around 1997 and immediately began looking for new fishing clients. In short order, he represented American Sportfishing Association, Nussman's outfit; The Billfish Foundation, Win Rockefeller, chairman, and Dr. Ellen Peel, president; and the Coastal Conservation Association (CCA). His real prize was the CCA, started in Houston, Texas, by Walter Fondren and other Texas sportsmen who were concerned that the redfish (channel bass or red drum) were disappearing from their inshore waters because of commercial exploitation. Paul Prud'homme, the famous chef in New Orleans, had created blackened redfish as a delicacy. Redfish landings soared and soon enough there was a problem with the stocks. CCA led the charge with the Texas Department of Fisheries and got laws passed to protect these fish. At the same time, a great sportsman and friend of mine, Harvey Weil, raised about $1 million, which was matched by Corpus Christi Power and Light to start an aquaculture program raising the redfish and releasing them into the waters in southwestern Texas. It took about four years, but the fish made a remarkable recovery. CCA then began a plan to create chapters in other states that were having similar inshore fishery management problems. They hit fertile territory. Red snappers in the Gulf of Mexico and Florida; seatrout of Florida and the Carolinas; redfish, bluefish, and striped bass of North Carolina and Virginia; and striped bass, bluefish, fluke, sea bass, and blackfish of Maryland, Delaware, New Jersey, New York, Connecticut, Rhode Island, and Massachusetts were all in trouble because of overfishing by commercial fleets either by shrimping or dragging. CCA informed its membership that for every one pound of shrimp they eat, nine pounds of juvenile game fish are killed, most of them thrown or shoveled overboard as a wasted bycatch. CCA hired lawyers to protect its membership's fishery rights and the organization grew accordingly.

Even NMFS recognized this problem. Many hearings were held, but when the problem fell back on the regional councils, especially the Northeast, Mid-Atlantic, Southeast, and Gulf councils, where the commercial interests outvoted anyone else by 16 to 5, 17 to 4, or 15 to 6; nothing got done and the waste and destruction to our inshore fishery stocks went on. Especially in the Gulf council, CCA led the fight to change the one-sided way of conducting the nation's fishery business, and adopted a pro-legal approach. The many lawsuits began to have an effect as decision after decision from the courts began to fall on the side of conservation and meaningful fishery management plans. Recreational fishermen knew a good fighter when they saw one in action, and by 2002, CCA had many active chapters.

Enter The Billfish Foundation and Dr. Ellen Peel, its new CEO. Ellen cut her ICCAT teeth on the 1997 billfish agreements. She is a secretive woman, not a team player nor a consensus builder. I know for a fact that several members of her board quit because she did not inform them of the MOU that she signed with Hayes, Nussman, Delaney, and Beideman who represented the longliners. This is odd because she is an attorney and has a doctorate in marine science. Her flaw is that she does not trust the people who are on her side as conservationists and champions of marine resources. She kept running out of the room at the 1997 ICCAT meetings, presumably to call someone in the United States for directions. One Billfish Foundation trustee from the Gulf told me that when she was hired he asked if she could work with Bob Hayes. She said, "no," because she did not trust him. Neither did the Billfish Foundation at that time. Consequently, in 1996, she was hired. Strange compatriots were to become the signers of the MOU (see Exhibit N at pages 243–247).

These unlikely players, (Nussman, Hayes, Peel, and Delaney, and Beideman of the longline's Blue Water Fishermen's Association) got together and formed a pact in June 2000. To

Delaney's credit he knew that NMFS was in the process of publishing in the *Federal Register* a proposed rule that would close certain areas of the Gulf of Mexico and Florida to longlining. He devised a scheme that would permit certain closed areas in exchange for a buyout program for unprofitable, older boats of the Blue Water Fisherman's Association's membership. He felt he had to do this in moderate secrecy and with no tinkering once an agreement was reached; that MOU provided that no change in the text could occur and that each signer would defend the memo against all other groups, thereby allowing no input from anyone except the original signers.

That they were all clients of Bob Hayes was not lost on the recreational community. Among other clauses, it provided for an eventual buyout of longline boats. This in and of itself was a great idea, but the price of $450,000 per vessel was outrageous because many of the vessels would not appraise over $75,000. This windfall for them was a rape of the taxpayer. Many of the longline vessels were non-steel boats, long on maintenance and short on capacity. These were the clunkers of the 250-vessel fleet. There was no provision for a charter or head boat buyout. Surely as commercial fishermen were hurt by the closures, the same was true of the charter and head boat owners. There was no provision regarding the providing of a tax return to show exactly what the earnings and worth or debt of the vessels were. The recreational community also felt that with proposed closures coming to the Gulf of Mexico and Florida, many longline boats might come, and had in fact threatened to come, to the mid-Atlantic states, thereby increasing fishing pressure in these areas. One year later, the repercussions are still being felt.

But one point is clear; the defeat of the Longline Five bill saved the American taxpayers at least $20 million.

A small consideration was a decline in almost every fishery under NMFS control. The shining star of the NMFS management firmament was striped bass. Here the commercial catch was very limited. Haul seining had been stopped, thanks to the unbelievable greed of the Bonaker families from eastern Long Island, and for four years recreational fishermen had put back every striped bass under 38 inches, millions of fish. The striped bass formula worked. The money in the longline buyout would have gone to the very people who had hurt the fisheries in the first place. Yet, the buyout was the only way to get fewer hooks in the water, and this meant less killing of all highly migratory species. But if you had some capital and wanted to work your own dragger or longliner, slim pickings were the order of the day, and here was a program, signed on by three major players in the recreational community, to effect a buyout of one-third of the longline fleet. Many of us working for the side of recreational fishermen felt outraged that a secret deal was made and made with a lawyer who a short time before had made his living representing commercial fishing groups. I also recognized the method used. If successful, the three groups Hayes represented could claim, and in fact did, that they were the saviors of billfish and many other highly migratory species like bluefin tunas. At stake were the potential twenty-five dollars in dues from sixteen million saltwater marine anglers in the United States. What could go wrong? Plenty.

CHAPTER 18

Office of Stephen Sloan
Telephone conversation with Jim Donofrio
(RFA) and Mike Leech (IGFA)
August 2000
1400 hours

• •

Jim Donofrio is the executive director of the Recreational
Fishing Alliance (RFA) based in New Gretna, New Jersey. His
office is part of the Viking Boat Works complex owned by the
Healey brothers of New Gretna. Jim was a former charter boat
captain, but got involved with the politics of fisheries at the
behest of the Healeys when the luxury boat tax was passed
under Bush 41's administration. The Viking owners were members of the National Marine Trade Association. In the early
1990s this organization recognized the reality that good fishing
or poor fishing would affect the ability to sell boats. Jim became
a registered lobbyist and is a fierce defender of recreational

fishing turf. He created the Recreational Fishing Alliance, which is now a nationwide group dedicated to conserving our fishery resources. Jim loves making political deals and has long tentacles into New Jersey congressmen, senators, and the state politicians. Mike Leech is the president of the International Game Fish Association (IGFA), the most easily recognized brand name in recreational fishing. It maintains a library of twenty thousand volumes, keeps the world's records on both freshwater and saltwater game fish, and is housed in a new museum in Dania Beach, Florida. I am a member of the advisory board of RFA and a trustee of IGFA.

All of us were flabbergasted at the news of the infamous longline Memorandum of Understanding (MOU). We felt that we had been stabbed in the back. No other recreational fishing organizations had been approached to sign such a memo. Hayes, Nussman, and Peel were extremely secretive about the contents. At the time of our telephone call, no one else had seen a copy other than the Longline Five—Delaney, Nussman, Hayes, Peel, and Beideman. The CCA monthly magazine and house oracle had just announced that the unpublished MOU was the greatest move toward the cooperation between recreational and commercial interests ever and implied that all other organizations, the non-signers, were asleep at the switch and that the CCA alliance should be given credit for saving the recreational community from an ocean disaster. The drive for CCA members was now at a high fever pitch. We were asked to believe, as an act of faith, that the MOU was good for everyone. I found this to be particularly galling because I thought that I had a strong relationship with Nussman through our work at ICCAT. I had told him when he was appointed that I would serve as a loyal lieutenant and never make an end run around him on any issues. Evidently, this understanding was not mutual. I never was informed of the MOU where the signers proclaimed that the buyout was the second coming of Christ, and the whole agreement was excellent for the preservation of billfish and other gamefish stocks.

We were told by the signers that it was history in the making and should sail through Congress because of its high moral tone and far-reaching fishery management concepts. "It was a new dawn of fishery management."

"Has anyone seen this document?" asked Donofrio.

"Nope," both Leech and I answered. "I understand that it calls for closing some areas of the Gulf of Mexico to longlining," said Leech.

"I heard that too," I chimed in, "also some parts of the Atlantic."

"But NMFS had already announced that they were planning a proposed rule, shortly to come out in the *Federal Register*, to close some portions of the Gulf of Mexico, and the Atlantic from Florida to the Carolinas; all this will do is send these boats off the Jersey coast," stated Jim.

"This is a very smart move by Glen Delaney and Beideman the Hammer. Knowing that they were going to have some sort of area closures prohibiting longlining sponsored by the NMFS, they have struck a deal to get paid for it as well. No one ever accused Glen and the Hammer of being stupid," I said.

"How much per boat?" asked Leech.

"I hear through the grapevine and some of Saxton's staffers that it could go as high as $450,000 per vessel. They will need a bill to go through Congress for this and I understand that a bill will be coming out shortly," Jimmy Donofrio chimed in.

Jim Saxton is a Republican congressman from RFA and Viking's district. The Blue Water Fisherman's Association and the Hammer were also in Saxton's district. Saxton, a decent honest man who at this time was the chairman of the House Fisheries Committee, had been closely associated with commercial fishing interests for years, but slowly and surely, through the efforts of the Viking yacht interests, including Jim Donofrio's efforts, he began to see the recreational sides of fishery issues. He listened and was fair in his dealings.

"You know what is clear to me," I said, "if you are not a client of Bob Hayes, you do not get to see anything and are the proverbial mushroom, kept in the dark and crapped on. How could this trio ever think they could negotiate anything of this magnitude without the support of all of the fishing groups or at least a vast majority of them?" I asked.

"This is a turf grab," said Donofrio. "It is all about membership, dues, and turf. They want to claim to speak for all recreational fishermen with this deal. I cannot get to work unless I see the memo and a copy of the draft of the bill. As soon as it hits Saxton's office I can get a copy. Where is the conservation here?" Jimmy asked, as did Saxton when he first read the memorandum. "I will bet that the closed areas they propose are areas that have no billfish at all. Then they will threaten us here in New Jersey with coming into our waters, because they have been precluded from fishing in the Gulf of Mexico and the Atlantic coast from Florida to the Carolinas." Jimmy's prediction proved to be deadly accurate.

"What I am steamed about is the secretive way they went about this memo," added Leech. "I understand that they have mutually agreed that they will not negotiate anything in the MOU. This is it, take it or leave it."

"We will see what Saxton and Senator Torricelli have to say about this. I understand that Senator Breaux is on their side and willing to sponsor the bill in the Senate. I guess this comes from Delaney's past association," added Donofrio.

"I will speak to Ken Hinman [from the National Coalition for Marine Conservation] and get his take on the matter. [I also serve on that board.] Why Nussman from ASA is mixing in on this instead of representing all of the industry and not just those three is beyond me. He has not shared this latest deal with any members of the ICCAT delegation, including me. You can take a staffer off the Hill but you cannot take the Hill's way of doing business out of a former staffer," I added. "This is going to end badly."

In ensuing weeks we finally got a copy of the proposed buy-out bill as drafted by Delaney, Hayes, and Nussman. It was a nightmare of loopholes and provided almost no conservation. It did provide for a huge buyout, at the taxpayer's expense, of some $450,000 per vessel without any accountability for getting the money. How about a tax return? How about proof that the vessel had fished in the past five years? None of this was in the bill. How about proof that at least 50 percent of the fisherman's wages came from fishing activities? I pointed out to my colleagues that to be "fair and equal" under the Magnuson-Stevens Act there should be a provision for some money for charter boats that were also going to be hurt by the lack of fish and stringent management measures taken by NMFS.

Congressman Saxton was not fooled one whit. "Where is the conservation in the proposed bill?" he asked almost immediately. As soon as Delaney, Nussman, Hayes, Peel, and Beideman heard Saxton's reply, they headed for the offices of Peter Goss, a congressman from Florida, to introduce another bill in the House in an attempt to circumvent Saxton's office. This is not a good strategic political move to try to navigate around the chairman of the House of Representatives Fisheries Committee. Saxton never said much publicly, but his staff was furious. A huge groundswell against the proposed buyout and area closure bill was building. Meanwhile, NMFS was on its merry way with its own version of a proposed rule for longline area closures. I must say that the NMFS version of the area to be closed and those proposed by the Longline Five were basically the same. None had any protection for the Mid-Atlantic, namely New York, New Jersey, Delaware, Maryland, and Virginia. Considering that New Jersey has the longest coastline, and Congressman and Chairman Saxton ran in a district closely aligned with ocean fishing, the Longline Five were sailing into the perfect maelstrom of ocean politics. The Longline Five's bill was short on accountability for the buyout and conservation for the nonprotected areas, but long on a secretiveness that spelled disaster for them. It raised the hope of some sixty-six longliners, who were expecting a windfall from the oversized buyout program, but it lacked the necessary political judgment from people who were inside-the-Beltway players (Delaney, Nussman, Hayes) who should have known better.

The harder the Longline Five tried to get the bill through the process, the more opposition they met. This led to a high pitch of frustration. Dr. Ellen Peel issued a famous directive, "It is the fish, stupid!" Calling the chairman of the House Fisheries Committee "stupid" was not politically adroit, to say the least, and I do not think that Saxton sent her a Christmas card. In early November, Delaney, in a moment of pique and frustration, let loose with a blast that certainly burned many political bridges behind him. Even so, the bill kept floundering and was finally shipwrecked when Congress adjourned in December 2000.

NMFS passed its own version of the closed areas and it is now the law. CCA and Hayes kept publishing weekly press releases calling RFA, IGFA, and NCMC (National Coalition for Marine Conservation) enemies of the state. Treason, sedition, and other fishing crimes and misdemeanors were mentioned just because some clear-thinking and concerned fishermen, and I include myself here, were questioning the motives and wanted some input into the process. Nussman privately admitted to me that if he had to do it all over again, he would have had twenty more signatures on the MOU. I agreed. It was hard enough to take on the commercial interests at ICCAT and their political clout without having to fight among ourselves. Greed, however, still is the most important commodity in any such negotiation. Nussman nodded his head in agreement when I expressed these thoughts. I still wonder if he meant it. When we met in Marrakech, Morocco, in November 2000, the bill, the drafts, the versions, the possibilities, the area closures proposed by NMFS, and the buyouts were all in play.

As of this writing, the proposed buyouts of the sixty-six vessels at $450,000 each is as dead as a swordfish on the icehouse floor of Tsukiji Market. This is not the case with the proposed marine protected areas (MPA). Environmental groups now want to section off large areas of the ocean and prohibit fishing of any kind, even hook-and-release recreational fishing. Of course, recreational fishermen trolling for gamefish only use the top 20 percent of the water column. They do not disturb the reefs and the bottom habitat.

The concept of restricting hook-and-release fishing, even in an area where only the top of the water column is used, leads to fighting words. The recreational community will never support this concept and the battlelines are rapidly being drawn. Perhaps it can be negotiated, but, more likely than not, it will be up to a federal judge. In my opinion, no marine protected area should be sanctioned without a complete ecological study. There are provable damages in the case for recreational fisherman and related industries should they be shut out of these areas. The final word is not yet heard in the year 2002. Both sides are playing out their positions in numerous articles in the press and in meetings and, most important, in the House of Representatives under bill number HR 4749 which is Wayne Gilchrist's version of the new Magnuson bill. Gilchrist is a Republican congressman from the first district in Maryland. He is chairman of the resources committee of the House. He previously had gotten high marks as a friend of recreational fishermen and the environmental committee. This is a fine perch until you are chairman, then the deals begin. Unfortunately he also has Don Young (R-Alaska), Billy Tauzin (R-Louisiana), and Walter Jones (R-North Carolina) on his committee. All three barely recognize there are recreational fishermen in the United States. All three dislike the environmental community and are doing everything in their power to neuter the reauthorization of the Magnuson Act to make it commercial user-friendly. Gilchrist wants to be chairman of the Resources Committee of the House and therefore is playing high stakes political poker with our nation's marine resources. That same uneasy feeling is on me once more. Jim Saxton, ride to the rescue.

CHAPTER 19

Holiday Inn
Islip, Long Island, New York
Turtle Scoping Meeting
August 2000
1950 hours

• •

I almost missed the NMFS notice of a scoping meeting called to discuss the new "Biological Opinion" regarding the interaction of longline vessels with turtles. It was published by NMFS in June 2000 and was a startling document. In one hundred pages it proved what many of us have suspected for many years. Longlining is a very destructive way of commercial fishing. One look at the table of contents reprinted in Exhibit O (pages 248–252) and one can immediately see why NMFS was concerned about the repercussions of the interaction of longline gear with loggerhead and leatherback turtles. Why the concern? It just so happens that both species are on the endangered

species list with CITIES (Committee for the International Trade
in Endangered Species). Both are rated as Appendix I cases.
Therefore no interaction, or handling of any kind, is permitted
under the CITIES agreements with member nations.

In the course of monitoring the longline fisheries, with even
a low percentage of observers, a high incidence of hooking these
endangered turtles was reported. (Exhibit O.) Frantically, NMFS
polled the commercial longline people for help in finding a way
not to interact and therefore not to catch these endangered turtles. No help was forthcoming because, basically, the quest was
futile. As long as the 547.5 million hooks were in the water every
year, the turtles would be caught in violation of CITIES and the
law. I met a former commercial fisherman at the August 2000
scoping meeting who told me the turtles are free-swimming in
the plankton-rich, jellyfish-laden waters where longliners ply
their trade. At night when the twenty or thirty miles of lines are
set with the 2,500 to 5,000 baited hooks, the turtles swim along
the surface and rub up against the balls, each the size of a soft-
ball, that are strung along the main longline every fifty feet or so,
keeping the ganglions where the hooks are attached dangling in
the water some sixty to a hundred and twenty feet below the surface. The turtle thinks the ball, painted orange, is a jellyfish and
nuzzles up to it and looks down. There in the depths below,
attached to the baited hook, is a Cyalume lightstick, which is a
plastic tube chemically treated inside to glow, illuminating the
bait in an eerie green or purple light. American Cyanamid makes
the "light sticks" and they are effective for attracting swordfish.
Unfortunately, other highly migratory species are attracted to
the illuminated baits. The turtles are not immune to the attrac-

tion and down they dive looking for a meal. There it is; a nice juicy squid or mackerel all lit up compliments of the light stick. The turtle gulps down the bait and soon feels the 10/0 circle hook in its throat. There is no escape. If the longline is heavy with fish, the turtle most often drowns. If the turtle has taken the bait close to haulback time, it may survive to the boat, but there is no disgorging that steel hook without a high mortality. If the turtle is cut loose with the hook embedded in its throat, it will starve to death and not survive. In addition, the law specified in the Endangered Species Act requires that no one touch the turtle to unhook it. This may seem draconian in substance, but it was enacted to stop the "incidental catch or take" of endangered species. It seems that the poachers and others who make a trade in endangered species, are ready, willing, and unfortunately, able to jump through the slightest crack or tiniest loophole. The CITIES contract with the signatory nations means that there is to be no interaction.

NMFS held the scoping meeting that August night looking for ways to ameliorate the situation, which the members themselves put into the "Biological Opinion." The threat to turtles from longliners is acute. Oddly enough, considering the stakes involved, there were only a dozen commercial longline fishermen at the meeting. They tried, as usual, to discredit NMFS's numbers. Then one suggested that the entire "Biological Opinion" be submitted to a "peer review" by the National Academy of Sciences. This is standard operating procedure. If you do not like what one report says, get another scientific group to reanalyze the report and try to get a different answer. One attendee went so far as to say there were plenty of turtles out there and so what if a few get killed. Phil Kozak from the National Fishing Association, a grassroots New Jersey-based organization, spoke to the attending longline fishermen and gave a spellbinding lesson on the whys and wherefores of the Endangered Species Act. He stressed that one is not allowed to touch a turtle, and perhaps the closures of areas where the interaction takes place would be necessary in order to get around the obvious conclusions from the "Biological Opinion" that longlining for swordfish and tunas is destructive to endangered turtle species populations of loggerheads, leatherbacks, and ridleys.

"Do you mean to tell me that NMFS may have to close all areas where longliners operate now?" asked one commercial fisherman, suddenly realizing what NMFS was suggesting in the "Biological Opinion." The affirmative answer he got back from NMFS personnel at the meeting really shook him up. In fact, NMFS people explained that they might be forced to do something by October 15, 2000, only two months away. This meant closing Georges Bank and the Grand Banks, among the most prolific areas for swordfishing in the world. The rest of the attendees in the room went white with fear. "This is serious shit!" one man exclaimed. I got up and made a pitch for the longline industry to figure out a new way of fishing for swordfish and tunas without interacting with turtles. There is one, I suggested, and it is called harpooning, which is basically selective hunting. It existed for many years, having had its heyday just before World War II and into the early 1960s. Longlining came along and it was the equivalent of the machine gun replacing the bow and arrow. More than one-half of America is two paychecks away from bankruptcy, and there was real fear in the audience that night of the scoping meeting.

As of this writing the issue is not resolved. NMFS has postponed several hearings regarding the protection of the loggerhead and leatherback turtles from longlining. The catch of these endangered animals goes on every day, not only in the United States fleet; but in all foreign fleets under ICCAT's aegis. In June 2001, NMFS issued its latest stall by recommending a large net be used for bringing the turtle on deck. This completely ignores the Act. Ho hum, another lawsuit will no doubt be filed by the environmental community compelling NMFS to close the areas where longlining is interacting with turtles. At this moment, there is a slightly positive note. NMFS held a meeting on March 12, 2002, entitled "Review of 2001 Pelagic Longline Sea Turtle Mitigation Research Results." The results and conclusions are found in Exhibit P at pages 253–256. It is interesting to note that it is almost two years as of this writing since the first scoping meeting described here. By now, millions of our taxpayer dollars have been spent desperately trying to find a way to let longlining continue, killing even more endangered turtles, all under the guise of science. All this effort to protect a $10 million swordfish business that has at the most one thousand jobs spread out from Maine to Texas. Cash-cash businesses are terribly hard to kill.

I have no doubt that the solution to getting the oceans partly under control lies at the feet of the turtle; as the Bible says, "And the voice of the turtle is heard in our land." It is now speaking, saying, "Save us, save us! and in doing so you will save yourselves."

CHAPTER 20

United States Delegation Room
Hotel Kenzi El Farah
Marrakech, Morocco
ICCAT meeting, first day
November 2000
1800 hours

• •

Before leaving the United States, Rich Ruais, head of the East Coast Tuna Association (ECTA) thought he had a new bluefin deal in his pocket, regarding an increase in quota of 200 to 250 metric tons (440,000 to 550,000 pounds), or another 1,100 to 1,375 four-hundred-pound fish for American fishermen, members of his association. Rich had done his homework before arriving at ICCAT's meeting in Marrakech in November 2000. NMFS received letters from Senators Kerry, Kennedy, and Snow, supporting his position. These letters went all the way up the NOAA chain of command to Dr. James Baker, the head of

NOAA. To reach Dr. Baker's desk, the pressures had to be a magnitude of a giant tsunami threatening to obliterate our naval base at Guantanamo. Rich Ruais fully expected to have a favorable decision because, in fact, the early science was on his side during the studies for the bluefin tuna's recovery plan. Recoveries of both regular and archival tags started to record bluefin tunas making their migrations from the Western Atlantic to the shores of Tripoli. In fact, three popped up off Sicily.

At the opening meeting of the American ICCAT delegation, the evening before the opening of the plenary session, Rollie Schmitten, our head commissioner, announced that he had spoken to Dr. Baker of NOAA the day he left for Morocco, and Baker had informed him that he could not go along with Ruais's request to negotiate for an additional 200 to 250 metric tons of fish. Baker was concerned that the request, coming only in the second year of a thirty-year recovery program, was premature. Rollie explained that Baker had allowed an increase in the New England groundfish quotas and the numbers that tumbled in after that were worse, therefore leaving NMFS and NOAA open to future lawsuits. The closing of the cod fishery off New England and in Canadian waters is a case in point. Dr. Baker had good reason to be concerned because NMFS today has the distinction of being the most sued governmental agency with some 113 lawsuits outstanding as of this writing. There are more coming at two to three a week. Basically the environmental and commercial fishing plaintiffs disagree with the NMFS's execution of the fishery management plans. NMFS had been losing case after case because it bowed to political and commercial fishermen's pressure to keep catch levels above a 50 percent chance of success, which is the criterion in fishery management plans. The federal judges are no dummies, and they immediately saw through methods and schemes designed to keep catch levels higher than required under the law and under published regulations and rules.

Ruais was flabbergasted at the NOAA turndown. Many times in the past at these meetings I saw him set his jaw and start to think his way out of a problem. I am sure he was grinding his teeth that night. The next day he met the two representatives from the Faeroe Islands, north of Scotland. They were not members of ICCAT but came to the meetings to monitor events. The Faeroes did not have any bluefin tuna allocation. In May 2000, I went to Madrid for an Intersessional ICCAT meeting called to examine "The Criteria for Allocations on Noncontracting Parties." Translated this means that one was not supposed to fish the Atlantic and Caribbean for bluefin tunas if one were not a member of ICCAT with specific species allocations. Such prospective members as the Faeroe Islands had stated in the past, "Give us some quota and we will join." The participating contracting parties with quotas responded, "Join and we will give you some quota." It is a rondo refrain that has never ended.

Predictably, at the May 2001 meeting in Brussels, we could not even agree on what the word "criteria" meant. In fish-speak, this means that you may want allocation for catching bluefins, but it is not coming out of my quota. This puts the nonaligned party in a Catch-22 situation (sorry, but the pun is too good to pass up). "I need a quota to fish legally, but no one wants to grant or give a share of his or her own quota." There is none available, according to the scientists from SCRS who have unequivocally stated that everyone should accept a reduction of quota (catch) to keep in line with the thirty-year recovery plan, which supposedly will maintain the fishery at maximum sustainable yield (MSY). The two quota numbers in the recovery plan are 28,000 metric tons (61.6 million pounds) for the EC including the Mediterranean, and 2,500 metric tons (5.5 million pounds) for the Western Atlantic (United States and Canada) including the Caribbean.

Rich Ruais asked the Faeroe representative how fishing was this year. "Great," was the answer. "We caught 24 metric tons [52,800 pounds] in our own EEZ and offloaded 142 metric tons of bluefins [312,400 pounds] at Saint-Pierre, a small island still owned by France off the coast of Newfoundland."

"But," sputtered Rich, "you have no Western Atlantic quota."

"Yes, that's true, and we do not have one from the east [the European Community] either."

"I am confused," said Rich, and he meant it. Here, with no approval or quota from ICCAT, an outside party caught fish and sold its illegal catch in the markets to Japan. Japan, meanwhile, had agreed not to buy any fish from countries fishing illegally. The Faeroes, a small group of islands, a self-governing unit within the Danish realm, defied the powers at ICCAT, made economic hay out of sawdust, and will keep doing this until ICCAT declares them to be an IUU party so as to prevent Japan from buying their fish. The loophole here is the phrase "declared an Illegal, Unregulated, Unreported (IUU) party by ICCAT." When illegal fishing by the Faeroes came up in the compliance meeting and Dave Balton from the U.S. State Department asked for a resolution declaring the Faeroes an IUU nation, he was turned down, led by the forces of Japan. Follow the money! The Faeroes's catch of 166 metric tons, equal to 365,200 pounds, was worth $2 million to $3 million on the dock and many millions more through Japan's 7 percent add-on system. Every time someone touches a tuna or parts of a tuna, at least 7 percent is added on and finally, when that bluefin reaches its final destination, it is worth $300 per pound. This is a form of quasi-legal piracy taking place on the high seas. The East Coast Tuna Association and the American bluefin tuna fishermen just saw almost their entire request for an increase in allocation, that they felt due to them for the stringent conservation measures they had practiced, highjacked by the Faeroes on the high seas. Worse, they were almost powerless to stop it, for the United States buys almost nothing from the Faeroe Islands, so any thought of trade sanctions was out of the question. As of this writing, a flash report came over the Internet that one bluefin tuna, weighing about four hundred pounds, sold in Japan for $172,000. This news will only increase the pressure from nonaligned countries to continue their IUU fishing; ICCAT or no ICCAT.

To compound the insult, the EC announced on the third day of the meetings that it had overfished its 1999 quota by 7,000 metric tons (15.4 million pounds), going from 28,000 back up to 35,000. Incredibly the EC stated that the agreement made at the last meeting, reconfirmed three times by Rollie Schmitten during the intervening year, was not correct and in fact, was in error. It claimed the "over and under" deal was never stated clearly, and would not be honored. The EC went back to the game being played before Rio in 1999. In addition, the "figures" on the kill of small tunas, those under five pounds in weight, were still not available by the third day of the Marrakech meeting. That Rich Ruais did not grind his teeth to powder was a miracle.

United States Delegation Room
Hotel Kenzi El Farah
Marrakech, Morocco
Third day ICCAT plenary session
November 2000
0900 hours

• •

I sat in the plenary session during the second day and listened intently while our State Department representative Dave
Balton tried his best to convince the fellow members of the
Permanent Working Group (PWG) to send letters of noncompli-
ance to Belize, the Faeroe Islands, Panama, and other countries
that were fishing out of compliance, that is, fishing without a
quota in defiance of ICCAT. Dave Balton is a pro and a credit to
the United States. He never goes into a meeting unprepared and
is a strong, calm negotiator. No matter how Dave presented the
facts, he was stonewalled by the European Community, playing

the beard for such members as France, Spain, Portugal, and Italy regarding the issuance of a warning letter, a formal demarche, or a recommendation declaring a country an IUU fishing party. In the background was Japan, secretly against sending the IUU notices. The ultimate weapon was IUU because once that had been approved and issued by the secretariat of ICCAT, upon the unanimous approval of the contracting parties, Japan had to agree to refuse to buy the fish from that offending country. No one at the table that day wanted any such letters issued, in spite of overwhelming evidence of IUU fishing. The EC's only statement sounded like a rondo from some obscure Elizabethan tune. "We must be evenhanded, and fair, fair and evenhanded, fair but evenhanded, evenhanded and fair, and fair and evenhanded." It was a carousel of the same words over and over again. It basically refused to put other nations on notice about IUU fishing. I watched Dave Balton's frustration mount and finally one of the IUU fishers spoke to the issue: "Give us quota and we will join ICCAT." To which the EC would again respond, "You join and then we will take up the issue of quota." Mainly they were talking about bluefin tuna and swordfish quota. Having attended the special session of ICCAT the previous May on the criteria of allocations, I could well understand the frustrations of the nonaligned countries, those that were not members.

The participants at the May meeting could not even agree on how the word "criteria" was to be interpreted. It was clear from that meeting that none of the contracting countries, those already members, was willing to give up any of its quota to accommodate those countries that wanted to join. To say that the nonaligned countries trusted those in power to give them quota for bluefin and swordfish would be disingenuous. So it went around and around. After two hours I was sure that ICCAT, as an international body, was on its last legs. The acrimony was too bitter, too deep, and the quota stances inflexible. To my amazement, the issue was put aside, as another coffee break was declared, which in effect produced a cooling-down period. The delegates retired to the open courtyard between the buildings nestled around a swimming pool where tourists sunned themselves, oblivious to the many pasty-looking men and women who were debating the future of the Atlantic Ocean's fishery stocks and future.

Members of our delegation met with Mr. Suzuki, chief scientist of the Japanese delegation, to talk about the SCRS scientific report showing a severe decline in white and blue marlin stocks. Our members met to discuss proposals both sides were going to make regarding billfish, especially white and blue marlin. The report the next day at our delegation's breakfast briefing was unbelievable. "Suzuki lost it!" was the report. He began yelling, then screaming at our people. "The SCRS report is wrong; it is fallacious. We are going to ask for a peer review. We will not accept it. And as long as your recreational fishermen keep killing billfish, expect nothing from Japan." Normally a reserved, calm scientist, the report of his actions was startling. In the cold light of reason, however, his outburst was to be expected. Blame the blameless, throw the spotlight on someone else, and shift the attention away from the perpetrator. I remarked that there must be some kind of horrendous pressure building up in the Japanese delegation. Rollie Schmitten, our leader and also the United States whaling and Atlantic salmon commissioner, pointed out to us and to the open session that Japan had a respected scientist at SCRS and he had signed off on the billfish catch reduction proposals to stop the decline of white and blue marlin stocks now classified as overexploited. Mr. Masanori Miyahara was Japan's new ICCAT commissioner, taking over from Mr. Nomura who spoke English well and was a man of wit and drive. Many times, from the seat of the chair of ICCAT's working group, Nomura overruled his own delegation to get things moving to satisfactory conclusions. He has now gone to the World Trade Organization in Rome. He will serve them well there. We speculated that perhaps the new head was unsure of his marching orders. We prepared for the worst. It came.

CHAPTER 22

United States Delegation Room
Hotel Kenzi El Farah
Marrakech, Morocco
Third day of ICCAT meeting
November 2000
1700 hours

• •

I love numbers and I relish analyzing numbers in real estate or business deals I am looking to invest in or buy. I delight in matching up numbers with concepts, especially in working with our federal tax code. I trained myself to do mortgage amortization tables in my head. The level debt mortgage or loan payment and its variations has made American businessmen and consumers alike economic powerhouses. That a buyer (borrower) can make the same level monthly payment for twenty-two years at 8 percent interest and 2 percent amortization (10 percent constant annual payment) and pay off his loan is a spellbinding concept. I absolutely get a chill when it all fits into place.

When Secretary of the Treasury Robert Rubin was at Goldman Sachs, he developed probability of risk theories by using equations and numbers in trying to take the risk out of trading currencies around the world. The Japanese have erected a statue to honor W. Edwards Deming, who showed them how to evaluate quality control of products on a production line. Ford Motor Company discharged him as some sort of crackpot, but his theories created the powerhouses of Toyota and Honda as well as many other Japanese businesses.

As I sat in the American delegation room the afternoon of the second day of the Marrakech meetings, I had a peculiar feeling. When one becomes facile with numbers, one gets a second sense that the numbers and the conclusions obtained from that data are correct or are flawed.

I sat there with my Compaq computer and opened it to a spreadsheet and began to make a list of the known constants of the Atlantic and Caribbean fisheries ICCAT covered in the North Atlantic Tuna Treaty, the basis of the organization. They were as follows:

What we know is the reported catches for all species; the total number of boats that longline, the total that bait fish, and the total of those that purse seine in each of the species tracked by the scientists of SCRS. We know the reported catches by the contracting parties year by year as reported to SCRS. I questioned our own delegation longline members and confirmed that the vessels fish 150 days a year. I confirmed the average landed or dock price of swordfish, bluefin tuna, and bigeye tuna. I confirmed that the average length of longlines laid out behind each boat averages between 30 and 100 miles and that the average number of hooks used in one set was 2,500. I did not know what is the efficiency of the hooks in the water. I could now make assumptions of the catch rate based on 2,500 hooks, 3,000 hooks, or 4,000 hooks. I could, based on all of my assumptions, determine the economic value of the realistic ICCAT catches. I could produce an Atlantic Ocean and Caribbean fishing profit and loss statement together with a balance sheet.

I felt I was getting close to proving the real ICCAT catch, not just the reported catch. The bottom line was astounding; at the lowest calculation taken: 1,460 longline vessels, averaging 2,500 hooks a day for 150 days produced the staggering sum of 547.5 million hooks a year just for the Atlantic Ocean and Caribbean waters. If one were to take a 5 percent efficiency, that being if one hook in twenty caught a fish, 27,375,000 fish were caught a year just by the longline industry alone. If that figure increased by just 1 percent, 32,850,000 fish were caught a year by the same methods. Therefore each 1 percent increase in efficiency amounted to 5,475,000 more fish killed a year. If one takes in the total by-catch possibilities, meaning those species not targeted as swordfish or tunas are targeted, I truly believe that one hook in ten catches such ocean fishes as sharks and marlins, dolphins and even turtles. Therefore, the same 547.5 million hooks catch 54,750,000 pelagic fish a year. At fifty pounds average per fish, this amounts to 2,737,500,000 pounds of pelagic fishes while ICCAT reports only 1,095,871,800 pounds. The true ICCAT catch is almost exactly two and a half times larger than whatever ICCAT reports to the scientists according to my business-approach calculations. I challenge the ICCAT community to dis-prove these figures. In fact, I will devote five radio programs (five hours of radio time) to the subject on my weekly radio show *The Fishing Zone*.

The total reported longline catch, species by species, to the SCRS scientists for 1999 was 500,992 metric tons (1,102,182,400 pounds). See Exhibit Q on page 258. What struck me immediately was if 57.5 million hooks went over the side a year, no one would put a hook in the water without a piece of bait, which was almost always a squid, therefore there was a massive squid kill just to support this industry.

It all started to make sense. For years I watched Spain and Japan skirt any discussion of high seas squid boats. In fact, both countries had made many forays into joint ventures in American waters to take the ilex squid. Squid are the basis of the food chain for highly migratory species. A lending officer at the New York Life Insurance Company once told me that "a joint venture is where one party starts with the expertise (say, Spain and Japan) and the other the money or capital (say, the United States). At the end of the venture their roles are reversed." It was certainly happening on the high seas in the Exclusive Economic Zones of the contracting parties.

I began to fill out the cells in my spreadsheet program on my computer. The formulas I put into the cells are in the supporting material in Exhibit Q. I began to put values on each of the main species; swordfish, bluefin tuna, and bigeye tuna. I then valued all other species at two dollars a pound at the dock. I also discovered that, using my formulas, the total catch was exceeded by 248,864 metric tons (547,500,800 pounds). This was 50 percent higher than was reported. I double-checked my assumptions. At 2,500 hooks average a night, let's say an average sixty-mile longline would have 316,800 feet in its length. That would mean one hook every 127 feet. I checked with members of the longline group in our delegation and found that this

was about the average, but, in fact, the hooks were set somewhat closer than that.

The conclusion from the numbers that tumbled out of the
formulas I put into my worksheet were that the real catch was
3.5 times the reported catch by the ICCAT countries. Therefore:

1. There was a massive amount of IUU fishing by everyone
   except the United States and Canada. But even the United
   States and Canada are susceptible to further quota cuts if
   the true reporting were received. Translated: The stocks
   are declining faster than even the scientists have assessed.

2. The by-catch was even higher than supposed and the sale
   of that by-catch helped to defray the costs of high seas
   fishing. No value was given to discards shoveled overboard as "uneconomic."

3. Subsidies given by the shipbuilding countries (Spain,
   Korea, and Japan) only increase the problem, not the profits. Overcapitalization of the fleets is as much a problem
   as overfishing; one is Siamese-twinned to the other.

4. Fully exploited levels of the highly migratory species covered by ICCAT were probably reached some four or five
   years ago. Therefore, we are fishing on borrowed time and
   biomasses. If the levels of exploitation discovered here
   are true, the species will severely drop below their ability
   to reproduce even with full moratoriums in place, which
   nobody will ever vote for at ICCAT. References to ten-,
   twenty-, thirty-year recovery plans are figments of a less-
   than-fertile imagination. If fished constantly below MSY,
   no species can ever come back or recover its previous lev-

els of the late 1970s. They will exist at new bottom fishing levels of a graph showing their true populations.

5. Not only were the parties to this ICCAT treaty (after all it was originally called the Commission for the Conservation of Atlantic Tunas) conducting a destructive catching program, but other forces were destroying essential habitat, and forage food as well. The bottom trawling (called dragging) industry in each of the parties' EEZ that use this method was destroying the habitat at an alarming rate. High seas squid jigging and seining were needed to bait every one of the 547.5 million hooks used each year.

If you had your money in a bank that was paying out 3.5 times the normal rate of interest, and eating into its capital 350 percent faster than it was created, you would pull your money out of the bank immediately. The Feds would shut the bank down, and that bank would no longer have access to capital markets and float new bonds or sell stock. The Atlantic Ocean and the Caribbean are that bank, and that bank is bankrupt today. The Indian Ocean and the Pacific have much of the same problems as well and cannot be far behind.

This is the new math I propose for the scientific community. It is much too difficult to assess the number of fish in the water. It is far easier to ascertain the fishing activity based on the probability of hooks in the water and the economic values of the catch. The red storm-warning flags should be flying from the highest peak at every meeting held on the world's fisheries.

Speaking of flags, there is a loophole a mile wide in all this process. It is called foreign flags of convenience. For example, a fishing vessel is built in Spain, owned by a Spanish company, and has a Spanish bank account where it deposits the proceeds of each catch. The vessel is fishing under the Spanish quota for bluefin tuna or swordfish. That vessel gets older, the insurance contract requires repairs, inspections, and upgrading. That vessel goes to a country whose flag is for hire and that does not have any such insurance requirements, say, Belize. For a $1,000 to $1,500 fee, the vessel can be reflagged in Belize. Let's say the vessel catches 3,000 metric tons of swordfish, flying the flag of Belize, and lands that catch in Abidjan, Ivory Coast. The fish hit the dock, the money gets wired to Spain, and the catch gets charged to whose quota? Ivory Coast, Spain, or Belize that has no ICCAT quota? The answer is Belize and it immediately becomes an IUU country. Belize did not fish for the swordfish and, other than the fee for the flag, got no economic benefit from that catch. Even worse is the fact that the Spanish company is then free to build or buy another vessel, fly the Spanish flag, and catch more swordfish under the Spanish ICCAT quota. This is double dipping at the expense of all reason, fairness, and the fishery stocks themselves. There is no other conclusion. All ICCAT scientific reports are undervalued by 25 to 50 percent or more.

I propose these solutions. Reflag every vessel fishing under an ICCAT quota with an ICCAT fishing flag tied to a registered number. No vessel could unload a catch without flying a legitimate ICCAT flag. For the memo on Brussels meeting see Exhibit R at page 259.

The adage of "red sky in the morning, sailors take warning," should be repeated at the beginning of each and every day, like the muezzins' first-light prayer at mosques. The media should be the muezzins. Constant publicity of the state of our oceans is

148

needed. We have so much to do and so little time. I kept thinking, as the numbers tumbled into place: "What does all this
mean in the short term? In the long run? At ICCAT? To the
underdeveloped countries that depend on the meager jobs and
the protein from the fish they keep and do not export? Who is
buying these fish? Who is making the money here?"

I will take them in order. What does this mean in the short
term? Remember that I am speaking for the Atlantic Ocean and
the Caribbean waters and only highly migratory species under
the jurisdiction of ICCAT. My numbers have nothing to do with
the many problems of the Northeastern United States and
Canada, where codfish and groundfish are in dire shape, or the
reef fish, snapper, and turtle problems of the Gulf of Mexico, or
other coastal conditions. Clearly, fishery management as prac-
ticed during the last decade is flawed. It is flawed because the
scientists, even with flawed data 3.5 times too low on the catch
side, have stated that stock assessments are below maximum
sustainable catch levels and have suggested large reductions in
catch or outright moratoriums. The reaction by the catching
countries in those situations is predictable. They scream like
hell for a peer review, claiming the conclusions are flawed. As I
have shown, they are, but not the way the ICCAT countries
would like us to believe. They are 350 percent too low.
Draconian measures are needed to recover the Atlantic and
Caribbean fishery stock to a mid-1970s level.

Just as the Japanese did on the first day of the session in
Marrakech when the billfish assessment came out for white and
blue marlin, this constant sniping at any published fishery figure that goes against catching and killing more stocks is now
standard operating procedure. No scientist can properly do his

job if he knows that his published opinion will be ridiculed and may cost him his job and reputation should he work for the fishery department of a catcher country. It is an untenable situation, yet in spite of this peer pressure, unbearable at times, the SCRS scientists have tried to "assess" the various stocks and still come to the conclusions that reductions in catch are in order to effect the fishery management plans proposed and voted upon by the ICCAT contracting parties. It is a weird dance that goes on.

Another method is to expand the recovery program in such a way that a ten-year plan becomes a thirty-year recovery plan. This kind of manipulation is ridiculous. Thirty years from today no one will be around to remember what the original plan was. I remember Kuzio Shima, Japan's head fishery minister, telling me that we do know about the ocean stock of highly migratory species. Somewhere out there in the vast oceans of the world, spawning may take place that will bring the stocks back. With the sophisticated electronics we have today, do not bet on this theory. We are learning more about killing these fishes than we are about saving them. The big-spawn theory is flawed and we should not count on such nebulous thinking regarding highly migratory fishery stocks. The United States got wise to this game and has asserted in its regulations that no fishery management plan can be proposed by NMFS unless it has a greater than 50 percent probability of succeeding in ten years.

Some of ICCAT's plans, for bluefin tuna for instance, go out thirty years. The program is revisited every two or three years, but what good is the revisit when the parties refuse to cut the mortality, even when proclaimed by the SCRS scientists? Rollie Schmitten, formerly head of NMFS and then head of the

International Desk on Fisheries at the Department of Commerce, undertook in 1996 the painful issue of compliance. In fact, the United States and, most times, Canada are the lone voices for compliance at ICCAT. Japan is for compliance, but only when it concerns Taiwan. The European Community, including Spain and France, is not for compliance. In fact, it has deliberately refused to honor the agreements regarding bluefin tuna and swordfish. In the short term, more and more countries have enlisted the aid of flags of convenience for their fishing vessels.

This practice must stop. In the short term, Japan must become a responsible citizen of the oceans. The two biggest fish markets in the world belong to major players at ICCAT. Japan runs the largest, Tsukiji. Spain's is second-largest and is in Madrid. By its own corporate admission, Heinz 57, a NYSE-listed company, is a one-billion-dollar player in the world's tuna markets. Next is Taiwan, and coming on strong is China. These four players are responsible for the majority of the Atlantic and Caribbean's short-term fishery problems. In the light of landings, effort, and vessels, the United States and Canada are relatively minor players and for the most part play by the rules.

There is no point in discussing the long-term view, because if we do not clean up our collective acts, there will be no long-term fishery to worry about.

For the underdeveloped countries, those that have an ocean coastline but no major fishery, and those countries that practice IUU fishing, I have some sympathy for their plight. ICCAT has asked them to join the fishing community, but will not give up any quota in place to accommodate their entry. The countries' argument has been, "We will join if you give us 'adequate' quota." South Africa has taken this to mean that, with or without quota, South Africa can make joint ventures with countries that have quotas to fish their rich waters, but at the same time go out and catch fish with no quota for itself. South Africa feels that as long as it reports the catches it makes with its own vessels, it is not responsible for other venture partners and their reporting methods. This practice has spread to other ICCAT parties such as Brazil. It is a fine kettle of scientific fish. The more this happens, the less confidence any ICCAT party has as to the numbers crunched out by the SCRS scientists. No wonder, when the reporting is from the ICCAT countries themselves. These are semantic games that are played among the fishing countries plying their nets and five hundred million hooks in the Atlantic and Caribbean per year.

In the Gulf of Guinea in West Africa, several countries like Senegal, Ivory Coast, and Ghana, have demanded and gotten the owners of the vessels to use one-half of the crew from the country in whose EEZ the vessel is fishing. These are paying jobs, one of the most valuable commodities in that area. I have discussed the enormous profits earned by purse seining in these areas. One look at the brochure of the Barreras shipyard in Vigo, Spain (Exhibit H), tells a story of building larger and larger vessels with more capacity. There are fishery management and stock assessments for all of the highly migratory ICCAT species, but there is no plan to limit fleet capitalization. In fact, fishing vessel shipbuilding has gone unchecked for the last twenty years.

If one were to follow the reasoning contained in my ICCAT ocean balance sheet, the conclusion is simple: In normal business or banking terms, do not build another vessel. There are too many now, the capacity is too high, and the fish are in short supply and cannot be reproduced quickly enough to make the operations profitable without massive subsidies. I began to look at the costs of doing business, and started with swordfish, which is a targeted species and has a separate quota at ICCAT. They are as follows:

Gross Income: The Catch:
25,550 metric tons = 56,210,000 lbs. @ $4.00 $224,840,000

Cost of fuel: $ 43,800,000
1,460 boats x 150 days @ 200 gallons a day.
(This accounts for refrigeration, generators,
power blocks, and the engine.)

Salaries:
1,460 vessels x 8 in crew @ $15,000 per year $ 175,200,000
1,460 vessels x 1 captain @ $30,000 per year 43,800,000

Hooks:
547,500,000 hooks per year; $ 5,475,000
10% loss; 54,750,000 @ $.10

Gear: $ 8,760,000
1,460 vessels @ 60 miles average equals:
87,600 miles @ 10 percent loss @ $1,000 each

Bait:
547,500,000 hooks @ $.25 each squid $ 136,875,000

Insurance:
1,460 vessels @ $15,000 each $ 21,900,000

Repairs & Maintenance: $ 17,520,000
1,460 vessels @ $12,000 each

Permits and fees:

1,460 vessels @ $500 $ 730,000

Dockside landing fees: $ 4,380,000
1,460 vessels x 3 trips @ $1,000

Food:
1,460 vessels x 150 days sailing @ $10.00   2,190,000

Ice:
1,460 vessels x $1,000 per trip @ 3 trips
per year $ 4,380,000

Bank Loans & Amortization: $ 58,400,000
1,460 vessels x $400,000 per vessel
x 10 percent

Total costs per year: $ 567,210,000

Catch Dock Value $ 224,840,000
Negative Cash Flow (Loss) per vessel $ (342.370,000)

This might logically explain the stampede to build more and more fishing vessels. But this is not the true figure and, in fact, it is a tantalizing fantasy. The would-be dream of every captain and many of the crew is, "If I could only scrape some money together I, too, would be able to own one of these vessels and live my life in comfort with a picket fence and a thatched cottage in some bucolic setting." Dream on, MacDuff! There is a reality check.

It came to me as I pondered these numbers in that conference room in Marrakech that something did not feel right about these calculations. I began to wonder about the productivity of the longlining effort. In fact, if a vessel laid out 2,500 hooks, how many of them actually caught fish? What would be the bottom

154

line on this productivity, for instance, if one hook in 100 caught fish? Would the boat be able to continue fishing? I thought, "No. What a waste. All that squid going for naught." I settled in at a 5 percent productivity rate. That means one hook in twenty would be productive. If that were the case, what would be the catch in the Atlantic and Caribbean under ICCAT jurisdiction? The word problem began to take shape at that very moment.

These new assumptions sent a shock through my system. I felt as if I had labored for the past seven years in some dark, dank cave mindlessly processing fisheries numbers almost as an automaton. Here was proof of what I had always suspected, but could never prove. There was a massive excess catch. Even worse, if one treats the Atlantic and Caribbean as one resource bank with its fishery stocks as the capital of that bank, these figures and data were most disturbing. We, all of us associated with ICCAT, were winking at numbers that made us feel comfortable at the plenary table, but not in virtual reality. We were, and have been, withdrawing deposits from this bank faster than we have been depositing the interest, that interest being the ability of fishery stock to reproduce and build up the balance once again.

Our theories of Maximum Sustainable Yield (catch) are flawed. If my calculations are accurate, and I have no reason to believe that they are not, we have been depleting, with all our SCRS science, the fishery stocks of the Atlantic and Caribbean, ICCAT territory, three to four times faster than reported by the scientists. This is alarming, because at this writing many of the species involved have already been reported as fully exploited, meaning they are below MSY and are in danger of not being able to recover unless there are strict moratoriums. If this is true, we will run out of HMS fish stocks three times faster than contemplated. We will also run out of squid at this rate. We will also create bankruptcies three times faster than expected as fishing vessels come back with lower and lower catches, compounded by the explosion in the number of fishing vessels at a time when the stocks are in a rapid state of decline. This will also put pressure on the insurance companies as claims will rise and profitability will decline. As claims rise, so will premiums. More and more vessels will try to obtain "flags of convenience" to skirt around the insurance regulations of their countries, and the risk to the fisherman on one of these vessels will increase to the point where he will be naked as to any meaningful coverage. This scenario will make a fisherman's job the most dangerous one on earth. If a ship sinks, so will the asset, and no claim will be recovered. I suspect that banks will be very reluctant to make future loans.

The ICCAT contracting parties reported for the fishing season 2000-01 a swordfish catch of 56,210,000 pounds (25,500 metric tons) for the Atlantic and Caribbean waters. At $4.00 per pound dockside, the total value of that catch is $224,840,000. The cost of getting that catch is $567,210,000, creating a negative cash flow to the catching entities of $342,370,000 per annum just for the swordfish catch alone. ICCAT reports confirmed by commercial sources that each landed swordfish averages eighty pounds; therefore, the ICCAT contracting parties and their scientists would have us believe that 702,625 swordfish were landed during the 2000-01 fishing season. Each fish, therefore, would have a value of $320, but each fish would cost $488 more than it brought as a landed price. This is a massive negative cash flow and surely would lead rapidly to any bankruptcy court in the normal course of corporate business. But, I was uneasy with these numbers even while I was willing to accept the concept that because of socioeconomic factors, the fishing fleets of the ICCAT parties were entitled to this huge subsidy, for it is not only for the swordfish; the subsidy would increase even further, factoring in similar subsidies for bluefin tuna (BFT), bigeye tuna (BET), skipjack tunas, yellowfin tuna (YFT), shark, marlin, and other small pelagic species. Something struck me as out of kilter. I decided to judge fishing productivity with attending economics and cash flows plus or minus. The word problem was quite simply as previously described.

Problem: If 1,460 vessels fish 150 days a year and set 2,500 hooks a day, how many baited hooks would be set each year?

Answer: 547,500,000

Problem: If 547,500,000 baited hooks are set each year with a squid, how many squid would be used?

Answer: 547,500,000

Problem: Has anyone considered this in any squid assessment?

Answer: I do not believe so.

Problem: If 547,500,000 hooks are set each year and only one hook in twenty catches a swordfish, how many swordfish would be caught each year?

Answer: 27,375,000

Problem: If 27,375,000 swordfish averaged 80 pounds each, how many pounds (metric tons) of swordfish would be caught each year?

Answer: 2,190,000,000 pounds (995,455 metric tons)

Problem: If ICCAT reported in 2000-01 that a total of 56,210,000 pounds (25,550 metric tons) of swordfish landed, or 702,625 fish, what would be the productivity rate of the longline gear?

Answer: One hook in 779, or .0128 percent

Problem: What would be the amount of overfishing and non-reporting of ICCAT parties if the hypothesis is true?

Answer: 38.95 times the reported swordfish catches

Problem: What is the economic value difference between the reported catch and the alleged unreported catch minus the cost of catching the swordfish?

Answer: $ 8,680,795,000

Problem: What then is the profit per vessel of 1,460 vessels?

Answer: $7,967,950,000 ÷ 1,460 = $5,457,500

It all started to make sense.

At sea swordfish and other highly migratory species are the international food and money chain, and therefore are targets at will of anyone who owns a boat and pays a modest fee to ICCAT. There is also another hidden cost to this fishing. As pointed out before, there are IUU vessels that do not even stoop to the mundane task of joining ICCAT. They are pirates, operating within a den of pirates. Remember, this is only one species, swordfish; the same methods can be applied for BFT, BET, YFT, and so on. It is up to ICCAT to investigate these numbers. I will deal with the viable solutions next.

CHAPTER 23

**Delta Flight 960
Madrid to JFK
November 2000
1200 hours**

• •

As I sat in my seat during this leg of the flight back from Morocco, I began to speculate just what my new calculations would mean in the world of ICCAT. Would it make a difference as to the way ICCAT's fishery managers looked at the Atlantic Ocean? How could I, almost as a lone individual, get this new theory to a point where it could make a difference? I was determined to write a book about my experiences. The tapestry of stories coming out of my experiences might get our government and other nations to revise their thinking and present methodologies. When I landed back in the United States I began searching for some scientific formula that might help to prove the numbers I had uncovered during the bouts with my mind and my computer. I stumbled on Poisson's Theory, or the Law of Large Numbers, or the Poisson Distribution for Rare Events.

For example, if a hook is set every hundred feet and the average number of hooks per set is 2,500 and these hooks are set a hundred and fifty times a year, what is the total probability of 100 percent of the hooks catching a fish? None. What about 50 percent, better; 25 percent, much better; 5 percent or one hook in every twenty? That sounds about right because any catch lower, say 2 percent, for example, would most certainly make the fishery uneconomic and unproductive. I decided that 5 percent was the bottom productivity of any number of hooks in the water per year. That means that one hook in twenty set would catch a fish, and if an average of 2,500 hooks were set, then 125 fish would be hooked and kept. By using the spreadsheet, and averaging the fish at seventy-five pounds each, that total catch is 200 percent larger than the reported catch under the ICCAT umbrella. Something has to be done. The Atlantic Ocean as well as other waters are in peril; it is in a fight for survival. It is not only the highly migratory species that have declined, but many inshore fisheries and their habitat have been decimated in a rush for fishery gold. Exposing the overfishing is, in and of itself, not enough. What are the solutions to this massive problem?

The Supply Side:

1. Each quota at ICCAT for bluefin tuna, swordfish, blue and white marlin, and any other fully or overexploited species as already determined by the SCRS ICCAT reports must be cut by 45 percent for a period of five years with no possibility of any increases granted because of "better reports."

2. Every country or political entity fishing for swordfish, bluefin tuna, BAYS tunas, or billfish must be granted entry into ICCAT. Each of the present member countries must give up 5 percent of its latest quota into a pool and non-aligned countries will be granted quota from that new pool. If a country does not accept the quota from that pool, it will be declared an IUU country by definition. Everyone should be under one tent.

3. Each country must invest in new aquaculture programs to make up for the loss in tonnage because of the 45 percent reduction for previously unreported overfishing. A recent case in point is the aquaculture of Atlantic salmon. The price is so low, because of the vast amounts of aquacultured fish, that it is uneconomic to fish on the high seas for wild stocks. The present high seas fleet has offered itself up for a five-year buyout with an accompanying moratorium of Atlantic salmon high seas fishery. This new news ought to put to rest some of the outrageous claims by the environmental community against fish farming.

4. ICCAT must become the clearinghouse for all sorts of Atlantic and Caribbean fishery products, including canning and aquaculture. It should maintain a balance between overexploited species and those reproduced by aquaculture methods.

5. ICCAT can become a stock market for fishery products caught in the Atlantic and Caribbean zone, very much like the commodity markets that operate now in Chicago. The ease of transporting perishable goods today makes this absolutely feasible.

6. All vessels fishing the ICCAT-EEZ areas must register and obtain an ICCAT flag. No fishing vessels flying flags of

convenience will be allowed to be offloaded anywhere in ICCAT-EEZ areas.

7. Each ICCAT vessel fishing in ICCAT-EEZ areas must register the tonnage capacity of the vessel.

8. For a period of ten years, no discards will be allowed. No high-grading will be allowed and therefore there will be no ocean waste. What you catch is what you keep. The ICCAT contracting parties must develop a sophisticated supply side marketing so there is no waste of an ocean's assets. Each vessel must declare its holding capacity and operate on the principle of FISI (First In Stays In). Marketing the now diverse products in the ship's holding tanks can be easily done with computers and the worldwide distribution network now in place.

9. No further longline, purse seine, or processing ships should be built for a period of five years without a corresponding reduction in tonnage from the destruction of older vessels. It is not enough that the older vessels go out of a fishery only to reappear again as vessels in some economically emerging country, thus encouraging the cycle of overfishing to begin once again. This is the case today.

Compliance:

1. See number 6 above.

2. No one can buy products from a fishing vessel that does not fly an ICCAT flag and is not registered with ICCAT.

3. Violations will cause a fishing vessel to be declared an illegal, unregulated, and unreported vessel (IUU), and therefore the catch cannot be sold or purchased.

4. Each vessel fishing in ICCAT-EEZ areas must have a vessel monitoring system VMS on board and it must be functional at all times.

5. ICCAT must maintain a "war room" that charts the position of every ICCAT flag-flying and registered vessel. All fishing destinations must be registered in advance and sanctioned by ICCAT. This is to prevent the quota's being reached in one place and fishing taking place in another for an already full quota.

6. A Chinese Wall should be built between the fishing bureaus and the scientists. No government should be able to "politically massage" the scientists and fishery managers because of perceived economic goals. We all admit that having no fish in the ocean is the worst scenario for the world's populations.

Financing Compliance and Fishing Quotas:

1. ICCAT could collect ten cents per pound for their activities as outlined herein.

2. According to its own published numbers, ten cents a pound is equal to a potential ICCAT budget of $200 million per year, fully adequate for the aforementioned goals to be accomplished.

3. ICCAT could organize and sponsor recreational fishing tournaments in countries like Brazil, Morocco, Spain, Portugal's Azores, and other countries that have excellent billfishing and bluefin tuna fishing. These international matches could be modeled after the Sharp Cup, once the most prestigious fishing tournament, held annually in Nova Scotia until the tuna runs stopped in the late 1960s. The total economic benefits could be enormous for a thousand-boat billfish tournament economic analysis. In this way, ICCAT could encourage the concept of ecotourism: a fisherman who does not kill his catch, by definition a recreational fisherman, is no different from a

person paying to watch whales, an elephant, or a lion. This concept will bring large cash flows of dollars to many countries, for tourism is the third largest industry in the world.

4. ICCAT must address the turtle issues, for these will not go away.

5. The number of hooks permitted in the ICCAT areas must be limited. Each permitted fishing vessel must state its hook capacity and usage. This may entail the limited use of longline gear. One hundred miles of longline gear at one set seems to be excessive and, perhaps, a limited use of small longlines should be established. We certainly need more ICCAT-budgeted observer reports in order to correctly assess this problem, and a minimum random coverage of 25 percent of the vessels should be maintained. It is conceivable that 100 percent observer coverage is needed if there is interaction with endangered turtles. All achievable out of the ten-cent tax on every pound of landed fish among ICCAT contracting parties.

6. The question is which is less destructive, and the answer is unknown and will only be known from observer reports over a five-year period.

The Demand Side

1. Greater stresses on the oceans and their fishery products will occur in the near term because of pressure from population growth and such shortages in alternative food supply caused by mad-cow disease, foot-and-mouth disease, and the like, and other such areas with medical or environmental problems.

2. All ICCAT countries should declare a voluntary no-fish-consumption day each week, and this should be world-

wide. This is easily done when the distributing countries
such as Japan, Spain, Taiwan, France, and others can
close their distribution places on that day. This would
reduce the demand by 14 percent.

3.   All ICCAT countries must go forward with aquaculture
programs for the fish-farm products. We must reduce our
reliance on ocean-caught fish until they have had time to
recover. The World Bank must become a part of this program so as to provide loans and capital. A put-and-take
contract system, that is, I grow and you buy, could be
established quite easily. In fact, these put-and-take contracts are easily financed.

4.   The environmental community must back off its opposition to fish-farming within a reasonable distance of the
coastlines and populations. The environmental community
has committed plenty of rhetoric against fish-farming, but
little scientific data. It is an issue, in my opinion, that it
wishes to reserve for fundraising. No one has proven that
if a few or even a lot of fish escaped from a pen it would
hurt any wild stocks of fish. The argument made, sur-
rounded by the hysteria, is specious. The farmed fish
came, perhaps many times removed, from a wild strain at
one time. There is a choice here: an ocean without fish
and an ocean habitat that is being destroyed for generations to come by dragging, trawling, and dredging or, on
the other hand, fish-farming that will add some fish fecal
matter to the water, but not destroy the habitat over wide
areas. I am absolutely convinced that some sort of recovery system of converting fish-farm fecal matter into fertilizer can be done. This will eliminate the concern of the
environmental community.

5.   Let us create an ICCAT fish inspection sticker that says
     this fish was farm-raised so we did not have to kill one in
     the ocean. Surely the public service media can support
     this one. The United States government can take some of
     the $4 million a year that it gives outright to the National
     Fishery Institute to tell us to eat fish and convert it to a
     program to inform us to eat farm-raised fish for a time
     until the oceans recover. With more than a half-billion
     hooks in the Atlantic and Caribbean waters each year,
     we have much to do and precious little time to correct
     our abuses.

CHAPTER 24

The Pew Oceans Commission, Testimony of Stephen Sloan On behalf of the American Sportfishing Association, The International Game Fish Association, and The Fisheries Defense Fund November 29, 2001

• •

The Honorable Governor Pataki, the Honorable Chairman Leon Panetta, distinguished panel, guests, and concerned citizens for the welfare of our oceans. My name is Stephen Sloan. I am chairman of the Fisheries Defense Fund and a member of the American Sportfishing Association. Recreational fishermen are deeply concerned about the health of our oceans. According to the National Marine Fisheries Service's recently published figures, 10.4 million recreational anglers spend over $20.7 billion each year. To quote Mike Nussman, the president of the

American Sportfishing Association, who testified before the Presidential Ocean Commission, "Of all the United States finfish landings, recreational anglers accounted for only 3 percent of that total measured in pounds of fish landed. But, in landing that 3 percent, recreational anglers spent over $20 billion. Meanwhile the other 97 percent of finfish landings by commercial fishing operations are valued at just $1.6 billion.

We have a large stake in the health of our oceans. One very much overlooked fact is that for the past thirty years, recreational fishermen did not wait until stocks of billfish and other species were in trouble. They voluntarily applied hook-and-release fishing from Maine to Texas, from California to Alaska, and convinced the governments of Mexico, Venezuela, and the Bahamas to do the same. There has been little acclaim for their conservation efforts, in fact, in a strange way it increased the pressure to kill more billfish as recreational fishermen discovered new grounds only to have them exploited by foreign fleets. Recreational fishermen are the only user group to have sustained a concerted effort by returning millions of fish to their natural habitat.

It is on this note that I wish to change hats and put on my official hat of United States delegate to ICCAT, where I have served my country for the past eight years. ICCAT is the International Commission for the Conservation of Atlantic Tunas (and other highly migratory species) under the North Atlantic Tuna Treaty, which was signed by thirty-one contracting parties (nation-states and the European Community). I also serve on the board of the International Game Fish Association, the National Coalition for Marine Conservation, and am an adjunct professor at the Rosensteil School of Marine and Atmospheric Sciences of the University of Miami as well as former chairman of MAFAC, the Marine Fisheries Advisory Council, chartered by Congress. My curriculum vitae [was] included with my written text and I am a nonpaid advocate for the survival of our oceans.

Rather than state that there is an insurmountable problem I am quite optimistic that the oceans can be saved, but only if we act quickly. More science, while needed, is not the complete answer. It is more compliance with the existing science and agreements already in place. I believe I have a solution. There are institutions in place that produce the data, that produce the science, that produce the quotas and allocations, that produce the rules and regulations. This is all done at ICCAT for the Atlantic, Mediterranean, and Caribbean waters. What then is going wrong? It is my opinion that no country, other than the United States and Canada, will vote against its own self-interest in fishery matters. If a stock is being overfished according to the scientists, and a reduction of quota is necessary, it just does not happen. All kinds of objections and obfuscations come into play. Typically stated:

1.   The science is flawed; the data is flawed;
2.   Loopholes: putting off new assessments while fishing on the old high allocations. This is the order of the day.

Then there is IUU fishing (illegal, unreported, and unregulated fishing). I would like to concentrate the balance of my remarks today. Japan has positively identified 344 IUU vessels fishing the high seas. Most of these were former Japanese vessels sold to Chinese Taiwan. These vessels have been reflagged. They bear flags from Belize, Honduras, Bolivia, and Panama. There is no control of reflagging by ICCAT. In fact, in November's ICCAT session, the Japanese admitted that IUU fishing is out of control and asked for help in this matter. The Japanese also admitted they continually buy from IUU vessels and, in a remarkable moment of mea culpa statements, admitted they were a part of the fish-laundering taking place on the high seas by their refrigerated boats as they buy from IUU vessels and non-IUU vessels alike, thereby "laundering the catches." The phrase, "laundering the catches," is a direct quotation. There are other problems associated with foreign flags of convenience vessels which I describe in the memo that is attached to this report as Exhibit R. Please consider this scenario: It is essential to eliminate illegal fishing with strict compliance while we do the science. This memo was distributed to the delegations of the United States, Canada, Japan, South Africa, the Philippines, and the ICCAT secretariat to date.

It occurred to me, much like the rainbow money to counter-attack the drug dealers, we need a clean sweep. Let me recommend the following:

1. All vessels fishing the high seas must be registered. Within the registration a landing capacity must be stated: How many metric tons can the vessel hold.
2. The vessels must fly a fishing flag, an ICCAT flag in the Atlantic and a equivalency flag for the Pacific and other waters. A registration number, equal to an offloading number, accompanies the ship. No port captain will be allowed legally to unload a vessel with whole, frozen, or parts of fish products without flying the flag.

170

3. The registration of the vessel, be it of Panama, Belize, or Spain, is only for ownership or insurance purposes, not for fishing purposes.

4. No country can buy products from any vessel not registered in the program. The cost to ICCAT of installing this system would be $2.5 million or $500,000 a year for five years. Surely the Pew Charitable Trust could syndicate this amount among the other concerned 501(c)3s and private citizens.

5. A VMS system must be installed in all high seas fishing vessels. The plan is simple and totally effective in stopping this piracy.

This year at ICCAT, I pointed out to our delegation that the purchase of the bigeye tuna from the IUU vessels by the Japanese refrigerated boats amounted to $750 million at cost. The eventual markups in the Japanese marketing systems as these fish passed through many hands was over $4 billion. A 1/10 of 1 percent ICCAT tax would amount to $4 million a year, enough to fund the entire budget of ICCAT, including compliance and the flag monitor system, plus a VMS monitoring system. In Maine we have a tertiary radar system, now mothballed, costing the American taxpayers hundreds of millions of dollars. We can reactivate this system and keep track of fishing vessels, with multinational help, all over the world. In my opinion one of the most important missions of your commission would be to bring various branches of government to help in this global task.

This is not just a NOAA project; it concerns every citizen of the United States and indeed, the world. As mad-cow disease broke out in Europe, and now Japan, new pressures will increase for edible protein. All eyes, nets, and hooks will turn to the seas, bays, and oceans. The ocean species under ICCAT are all, repeat all, exploited, fully exploited, or overexploited. In the macro sense, this commission can get the government's other agencies like the Department of Defense with our undersea listening devices to help in the enforcement and compliance of rules for protected fishing areas.

In the post-September 11, 2001, era, there is excellent documentation that drugs and arms and possibly weapons of mass destruction are routinely transported on foreign flag-of-convenience vessels. If we start with my proposal regarding the reflagging of fishing vessels with a fishing flag, we have begun a process that can only help citizens everywhere learn that there are threats of terrorism as well as the health of the oceans themselves. What's in a flag? As far the oceans are concerned, everything.

CHAPTER 25

Hotel Meliá 7 Coronas
Murcia, Spain
ICCAT Plenary Session
November 2001
0900 hours

● ●

I sat in the plenary session listening to the interpreter drone on about the French purse seine fleet in the Mediterranean and the Gulf of Guinea. Then the speaker was Toru Furuhata, head commissioner from Japan. His second sentence hit me like a lightning bolt. "We here in Japan have been guilty of fish-laundering," he said in a flat monotone. I could not believe what I had just heard. "Fish-laundering." I wonder what he means by that, I asked myself. The next several sentences were punctuated by the same phrase, "fish-laundering." Normally a member of the contracting party delegation leaves a written document of his position in the cubbyholes set up for all attending the meeting with proper standing. There was no document in my box that morning. The delegate from Japan said that we all would have a document by noon. He gave no further explanation for the use of this unusual expression. I received the document late in that afternoon, and it contained an extraordinary concept. Fish-laundering was thus described by the Japanese:

There are three types of longline vessels involved in "fish-laundering."

The first group are the IUU (illegal, unregulated, and unreported vessels), some 342 in number. The second group is owned by business interests in mainland China, and the third is owned by individuals and business interests in Chinese Taiwan. All three fish the high seas in ICCAT jurisdictions (Atlantic and Caribbean). All three ownerships are serviced by refrigerated or reefer boats that are 100 percent owned by Japanese interests. These boats roam the oceans and collect the catches from the longline vessels owned by these three entities. This saves time and money, because the longline vessel does not have to go back to port with its catch. The owners of the reefer boats then sell their catch to another entity 100 percent owned by Japanese business interests that import/export the fish to Japan or sell the catch through other markets developed in Hong Kong or Africa or Spain. The "laundering" comes into play when the reefer boat collects catches from an IUU vessel and mixes them with catches from other boats. The body of the fish, called the plug, is indistinguishable one from another when covered with frost or held in a brine mixture in the hold of the reefer boat. Japan identified 342 IUU vessels fishing the ICCAT-EEZ areas. The dollar value of that catch, solely from these IUU vessels, per

174

year is $119.7 million. This figure is arrived at by multi-plying $350,000, the average dollar value catch by a longline vessel (Japan's own numbers) by 342, the number of IUU vessels reported by Japan. A complete list including names of the vessels was given to the ICCAT contracting parties. It would be a fair assumption that the final markup on the fish so "laundered" by the Japanese reefer boats and sold again and again within the system would be raised by a factor of ten to $1.197 billion.

Later that day in the American delegation room, I drew a chart showing the colossal amount of money involved, and I made a statement that, since Richard Nixon's speech about his dog Checkers, this is the greatest mea culpa statement ever. Japan is saying, "Look, we are guilty of fish-laundering, mea culpa; we do not know how to stop this practice, mea culpa; but since we are doing it, will you all at ICCAT please help us to get this practice under control." Mea culpa once more. This request came from Japan, the greatest fish profiteer known to history to ICCAT, an organization that has never brought one species back to health and could not even agree on the meaning of the word "criteria" when the allocation meetings were held in Brussels in May 2001. What is the likelihood of ICCAT's doing anything to monitor or put some enforcement into the concept of fish-laundering? Slim or nil, I would say. Perhaps it will be years before anything is done, and the IUU fishing goes on and on. Japan called for a conference in May of 2002, in Tokyo, to discuss IUU fishing. No meaningful measures for control of IUU fishing came out of the meeting.

CHAPTER 26

Recommendations

• •

1. ICCAT issue its own fishing flag.
2. All CCP and COOP must register their vessels with ICCAT.
3. The information shall contain tonnage catch capacity and document number.
4. The present flag of convenience the vessel is flying and a statement declaring to whom the catch belongs.
5. A coordination of F/V logbook with the ports of entry for offloading of catch.
6. Two warnings and then the third is automatic IUU declaration against the vessels, the FOC, and the country.
7. All fishing vessels must declare ownership. Example: A F/V from Taiwan, joint venturing with a Brazilian fishing company, must state the vessel is Taiwanese, but the catch is allocated to Brazil and must fly ICCAT flag for offloading purposes.
8. All factory processing vessels (FPV) must fly the ICCAT flag and declare the origin of the catches before offloading at a port.

9. The ICCAT flag will take precedence over any other flag flown by a F/V while offloading a catch.
10. All port captains will be notified of the ICCAT flag and its regulations.
11. Any violation of the port rules will subject that country to being declared an IUU violator and then sanctions will be issued against the unloading of future catches.

Presented to the ICCAT Secretariat, the delegation of Japan, the delegation of South Africa, and the delegation of the United States in Brussels, May 2001.

Final Thoughts in Late 2002

The United States now has the Pew Oceans Commission, and the Presidential Ocean Commission conducting hearings to prepare a report to the President and Congress about the state of our oceans. These reports are expected by January of 2003. Both commissions are well-intentioned, but to date I have heard no definitive word of suggestions about compliance. Do we dare to impose trade sanctions on fishery violators while we are dependent on other countries to help us in the fight against terrorism? I think not. Therefore we must concentrate in areas where the average citizen of any country can make a difference; that being the supply side. We, the people, eat seafood. We must be the ones who become product ocean-sensitive. Before you buy that Toyota, Honda, Lexus or Sony product, think twice. Before you order swordfish, Chilean sea bass, or marlin on the menu, think about all the undersized swordfish that are killed before they have had a chance to spawn at least once.

Politically, I see more of the same. The ICCAT and other ocean apparatuses are too big and ingrained in loophole-itis to be effective on the compliance side; too greedy to be effective on the conservation side; too untrustworthy to believe in the data side. As an aware citizen I see a great crusade. Just remember how the great fishery game is played. We can effect the measures I suggest in this book, but it will take an iron will of the environmental community, the recreational fishing community, and the concerned consuming public. Let this book be but a first step.

Afterword: A Friend of Fish

A few years back, New York City launched an intensive anti-littering campaign with a simple, straightforward TV ad: the camera followed a man walking down the street, unwrapping a candy bar and then tossing the crumpled wrapper in the direction of a litter basket. The wrapper misses and blows down the street as these words appear on screen: "When You Throw It Away, Where Is 'Away'?"

Among the people who know the answer to that question or its alternative, "Where is 'away' when you wash it away?" is my good friend Stephen Sloan. Steve, and a few others, but not nearly enough, know that "away" is the ocean. Everything is connected, everything flows downhill and therefore, Steve reasons, just about everything ends up in the ocean. This simple fact accounts for much of the alarm over the health, or lack of it, of the world's ocean systems. Too much stuff—everything from sewage, industrial and agricultural runoff, garbage, silt from such rivers as the Mississippi—winds up in the ocean and there floats, dissolves, settles to the bottom or washes up on shore in the form of oily sludge, plastic bottles and bags, tampon dispensers, condoms, abandoned gill nets, gruesome medical waste. (A few years ago, bathers in the exclusive Hamptons on Long Island were greeted by sopping wads of bandages, used syringes, even human viscera left over from surgical procedures.)

Human assault on the oceans and their inshore estuary systems is unabated, even accelerating: The Chesapeake Bay, perhaps the world's largest and most complex marine "nursery," used to have enough oysters in its creeks and bars to filter the entire body of water every 36 to 48 hours, keeping it clear. Today, it would take over 400 days to accomplish a favor nature provided at no cost, and the Chesapeake's water resembles *café au lait;* all due to the intrusion of human activity without appropriate planning, buffering, or simple concern.

Yet we, as a species, blithely assume that the oceans will both absorb an endless amount of discharge and abuse, and provide a boundless harvest of food and other resources. We don't understand that by draining the salt marshes, building on the dunes, or choking the estuaries with yet another marina, we are fiddling with the balance of the ocean, crippling its capacity to sustain not just its own natural chain of being but, ultimately, our way of life.

Standing on the ocean's shore, we let our gaze sweep over the water's surface and we marvel at its sparkling beauty, or cower in awe at its storm-driven fury. But because we cannot easily look *below* the surface, we have only a small understanding of what is going on down below—which is simultaneously too much and not enough.

Stephen Sloan's book, backed by decades of tireless (and voluntary) service on the boards of a dozen conservation groups and by countless hours of serious study, provides an informed, balanced, and compelling argument *pro mare.* It unlocks the mysterious workings going on *under* the surface, those systems and processes vital to sustain the ocean's resources. Steve explains and illustrates the constant assault on the world's oceans, too often fostered by greed and corruption. He debunks the myths that the ocean can satisfy our growing consumer needs, that its ability to provide "harvest" is infinite.

By being in it not for profit or self-interest, Steve is able to be an honest advocate for the ocean habitats and for the marine creatures most of us take (and eat) for granted. Too many lawmakers, regulators, recreational and commercial interests, even scientists, have an agenda, sometimes cleverly disguised as being for the public good. Too often, their eyes are on a larger slice of the ocean's pie. Too often, the public, uninformed, is simply blind to the truth, that the ocean systems are strained well beyond their carrying capacity, in some instances, beyond even mitigated (as opposed to natural) recovery.

Read this book and take its blunt lessons to heart. Then ask yourself, "What can I do to help?" Plot a course of action based on the kind of involvement that supports Stephen Sloan's reportage and commonsense solutions. Get involved. Do something for the ocean's health. Our future depends on it.

Richard Reagan,
President
Norcross Wildlife Foundation

EXHIBITS

INTERNATIONAL COMMISSION FOR THE CONSERVATION OF ATLANTIC TUNAS

CONTRACTING PARTIES
(as of December 31, 2000)

Angola, Barbados, Brazil, Canada, Cape Verde, China, Côte d'Ivoire, Croatia, Equatorial Guinea, European Community, France (St. Pierre & Miquelon), Gabon, Ghana, Guinea Conakry, Japan, Korea (Rep.), Libya, Morocco, Namibia, Panama, Russia, Sao Tomé & Principe, South Africa, Trinidad & Tobago, Tunisia, United Kingdom (Overseas Territories), United States, Uruguay, Venezuela.

COMMISSION OFFICERS

Commission Chairman	*First Vice-Chairman*	*Second Vice-Chairman*
I. NOMURA, Japan (22 November 1999 though 31 March 2000) J. BARAÑANO, EC-Spain (Acting, since 1 April 2000)	J. BARAÑANO, EC-Spain (since 22 November 1999)	A. SROUR, Morocco (since 22 November 1999)

Panel No.	*PANEL MEMBERSHIP*	*Chair*
-1- Tropical tunas	Angola, Brazil, Canada, Cape Verde, China, Cote d'Ivoire, European Community, Gabon, Ghana, Japan, Korea (Rep.), Libya, Morocco, Namibia, Panama, Russia, Sao Tome & Principe, Trinidad & Tobago, United Kingdom (Overseas Territories), United States. Venezuela	Cape Verde
-2- Temperate tunas, North	Canada, China, Croatia, European Community, France (St. Pierre & Miquelon), Japan, Libya, Morocco, Panama, Tunisia, United Kingdom (Overseas Territories), United States	European Community
-3- Temperate tunas, South	European Community, Japan, Korea (Rep.), Namibia, South Africa, United Kingdom (Overseas Territories), United States	South Africa
-4- Other species	Angola, Brazil, Canada, China, European Community, Japan, Morocco, Namibia, South Africa, Trinidad & Tobago, United Kingdom (Overseas Territories), United States. Uruguay, Venezuela	United States

SUBSIDIARY BODIES OF THE COMMISSION

	Chairman
STANDING COMMITTEE ON FINANCE & ADMINISTRATION (STACFAD)	J. JONES, Canada (since 21 November 1997)
STANDING COMMITTEE ON RESEARCH & STATISTICS (SCRS) Sub-Committee on Statistics: S. TURNER (United States), Coordinator Sub-Committee on Environment: J.M. FROMENTIN (EC-France), Coordinator Sub-Committee on By-catches: H. NAKANO (Japan), Coordinator	J. E. POWERS, United States (since 24 October 1997)
CONSERVATION & MANAGEMENT MEASURES COMPLIANCE COMMITTEE	J. F. PULVENIS (Venezuela) (since 22 November 1999)
PERMANENT WORKING GROUP FOR THE IMPROVEMENT OF ICCAT STATISTICS AND CONSERVATION MEASURES (PWG)	E. PENAS (EC) (since 22 November 1999)

ICCAT SECRETARIAT

Executive Secretary: Dr. A. RIBEIRO LIMA
Assistant Executive Secretary: Dr. P. M. MIYAKE
Address: C/Corazón de María 8, Madrid 28002 (Spain)
Internet: http://www.iccat.es. *E-mail:* info@iccat.es

Exhibit A

Organization and Structure of International Commission for the Conservation of Atlantic Tunas

by
STEPHEN SLOAN

On August 11, 1993 at the SW Corner of the Dumping Grounds (loran coordinates 14400/43600) the purse seiner "CONNIE JEAN" took a set on approximately 50 metric tons of bluefin tuna, all in the 100- to 200-lb. class. This set was witnessed by several charterboats in the area, and by others who took video tapes of the incident. The set was illegal as the season was to open on August 15, 1993. Calls started coming in to my office almost immediately. Charter captains reported that fishing in the area had been excellent the week before, with anglers and charterboats doing exceptionally well while trolling. Reports after the set showed that the bluefin tuna catch came to a complete standstill.

Much to everyone's surprise, a NMFS observer was on board. I interviewed Ms. Patricia Gerrior, a supervisor in the Federal Observer Program and the observer of record aboard the "Connie Jean" at the time of the incident. The interview took place via telephone on August 15th, four days after the incident and one day after the vessel returned to port. Several of her comments to me were, "The crew was unorganized and confused." "The crew was astonished that they were bluefins." "When questioned why the net remained in the water for over an

hour with fish, as many witnesses had reported, her comment was,"I don't remember the net being in the water for over an hour."

The question now becomes obvious. Can a tuna, who must swim to live, survive being held in a closely packed net for over one hour?

Why has this incident aroused and inflamed recreational fishermen throughout the entire East coast? It is because we all suspected that incidents like this happened continually through the years, but, no one was ever around to film them, until now.

Time after time, recreational and charterboat fisherman have called NMFS, and then our Conferderation offices, to complain about great fishing until the planes and purse seiners arrive on scene. The pattern over the last few years seems to indicate that just when the sportfishing starts to improve and captains and their charters are happy, in come the purse seiners and the bite quickly comes to an end.

The science behind all of this is well documented, just look at the ICCAT reports. In Madrid last year, the stocks of yellowfin, bigeyes, and skip jack for the very first time have come under alarming reduction.

The purse seiners are capable of massive sets, taking 100

Exhibit B
"The *Connie Jean* Incident"

186

The "Connie Jean" Incident

tons at a time. No fishery can take this heavy pressure. These purse seine vessels are very efficient. The highly migratory stocks of tuna cannot continue to be depleted at the present rate without a total collapse of the fishery. Look at the recovery of the striped bass once the haul seining was stopped.

Don Aldeus, a Canadian Government Fishery official informed me that we too could have fantastic tuna fishing off our shores if we put a stop to purse seining, as Canada has done.

In addition to the video tapes tapes in our possesion of the August 11, 1993 set made by the "CONNIE JEAN", I also have video tapes of other sets made on August 12, and August 13 by the same vessel. We are now checking to see if the August 11 set was the only one that bluefin tuna were involved.

What is wrong here? Why have five vessels capturing, with our governments blessing, 28-percent of the bluefin tuna quota, when 9,500 fisherman have eight percent of that quota.

The commercial catches are translated into money—billions. The recreational side is translated into use, trips and pounds of catch. Not one mention of dollar values. And here's the rub: Nowhere, and believe me I have searched high and low, is there a federal report in NMFS showing the socio-economic value of the recreational/charterboat industries to the coastal states. I feel this is by design and is a deliberate attempt by NMFS to undermine the recreational anglers cut of the proverbial pie.

It is further fueled by the fact that the S-K (Saltenstal - Kennedy) Grants, totalling some $60 million, which are doled out by NMFS to aid our fishery, come from a tax on the import of fish products. Therefore, who is entitled to this money, the commercial industry and a cadre of consultants, scientists and report producers or the recreational/charterboat fleet? There is a $1 million grant given to the commercial industry to tell us that fish are good to eat, how much did the recreational/charterboat fleet actually get from this honey pot? One hundred and twenty thousand dollars, of which $86,000 went to the IGFA to help with library indexing.

We, the largest user of these resources, get a pittance and are denied that very ability to produce a report that will show how meaningful our industry is to the economic welfare of the Atlantic Coastal States. I know that the State of New Jersey has a prepared a report that shows the economic impact of their recreational fishery, but the problem is just that; it is New Jersey's report and not in the loop as far as the feds are concerned.

A libel lawsuit has been currently filed by the owners of the "CONNIE JEAN" against our organization. And so, we have just been handed a golden opportunity to get justice and our fair share of the tuna fishery. We intended to countersue for $12, 500,000 for damages done to our fishery during this incident. We will be asking for your financial support to gain control of our fair share of this public resource.

If your club wants to see the films we currently have in our possesion, please write me on your club stationery with your available dates.

Join the team! We are about to be counted in a big way. You can help by contributing the cost of a flat of butterfish to the Fisheries Defense Fund. If you are really upset about our fishery, make it two flats ($50). Send your contributions to the Fisheries Defense Fund, Inc. 230 Park Ave, Suite 1221, New York, NY 10169. The time has come to fight back!

Exhibit B
"The *Connie Jean* Incident"

187

AUGUST 1993

VOLUME 5

FISHERIES DEFENSE FUND

50 TONS OF BLUEFINS DUMPED ON AUG.11

STEPHEN SLOAN CHM.
212.867 3730

PURSE SEINER CONNIE JEAN

SW CORNER OF DUMP CONNIE JEAN DUMPS AT LEAST 1000 BLUEFINS TO THE BOTTOM.

NMFS OBSERVOR ON BOARD-CLAIMS LITTLE KILL. CAACC MEMBERS ON SCENE REPORT OTHERWISE.

NMFS CLAIMS CREW INEXPERIENCED.

ENFORCEMENT OFFICERS ARE INVESTIGATING.

3% OF US QUOTA

100 TONS OF YELLOWFIN TAKEN DAY BEFORE BY SEINER ELIEEN MARIE

3% OF US QUOTA MAY HAVE BEEN KILLED IN ONE SET BY SEINER CONNIE JEAN

NEED FOR OBSERVORS

US BLUEFIN FISHERY UNDER SEIGE BY SEINERS LONG LINERS PAIR TRAWLERS

186 GIANTS TAKEN IN 2 DAYS OFF CHATHAM MA. BY SEINERS.

Exhibit B
Fisheries Defense Fund Broadsides

NOVEMBER 1993 VOLUME 7

FISHERIES DEFENSE FUND

LIBEL SUIT BEGUN BY
CONNIE JEAN & PILOT

STEPHEN SLOAN CHM.
212.867 3730

THE TRUTH WILL OUT!!!

CLAIM IS DAMAGE TO REPUTATION OF OWNER INGRANDE AND PILOT HILLHOUSE.

SUES: AL ANDERSON., THE FISHERMAN MAG. CAACC , FISHERIES DEFENSE FUND & STEPHEN SLOAN.

SLAP SUIT INTENDED TO MUZZLE FISHERMEN AGAINST PURSE SEINING.

ONE FLAT HELPS!!!!!

ALL DOCUMENTS NOW AVAILABLE TO OUR CAUSE!!

IF EVERYONE GIVES THE COST OF A FLAT OF BUTTERFISH $25.00/$25.00 : WE CAN WIN OUR FISHERY BACK FROM SEINERS!!!!!!!!!!!

THIS IS OUR CHANCE

BOSTON ATTY RETAINED: BINGHAM, DANA & GOULD BOSTON GLOBE REPORTS ON INCIDENT

CASE MOVED FROM N.BEDFORD TO FEDERAL COURT BOSTON

Exhibit B

Fisheries Defense Fund Broadsides

Tokyo's Tsukiji fish market is one of the largest in the world. How many billfish are the Japanese turning into sashimi each year? No one knows.

Exhibit C
Billfish Plugs in Tsukiji Market

The MARLIN WARS

Are the oceans being turned into deserts? The arrest of a commercial fishing boat in Mexico points up a world-wide problem. Bill Pike reports.

t was late February. The Mexican gunboat from Cabo San Lucas was closing rapidly with the *Copemapro V*, a rust-streaked 250-foot longliner wallowing in the long swells of the Pacific. The captain of the gunboat, Alejandro Salomon Velmar, studied the longliner through a pair of heavy binoculars. It appeared to him that the ship was breaking the law—working deep within the so-called "50-mile zone," a coastal area supposedly set aside for sportfishing and protected by the Mexican government from commercial exploitation.

As soon as Captain Velmar ordered the *Copemapro V* to heave-to, his VHF began to spew forth a babel of excited Spanish, Japanese, and English. It is quite reasonable to assume that the career Coast Guard officer's mouth was a little dry, his forehead filmed with a light, nervous sweat. What he was about to do, Captain Velmar must surely have suspected, might very well have far-reaching consequences.

Indeed, within a matter of weeks, a much-publicized scandal would result, separating three, highly-placed Mexican Department of Fisheries officials from their jobs with the slick efficiency of a filleting knife. Astronomical fines—the heaviest of their kind in Mexican history—would be imposed and the longlining permits of the biggest fishing syndicate in Ensenada—Copemapro S.A. de C.V.—would be revoked.

It would be a denouement quite unprecedented in the annals of the so-called "Marlin Wars," an only half-ironic appellation invented by charterboat people and hoteliers of the Baja to describe more than a decade of feuding over the industrial-strength harvest of billfish in local waters by the Japanese.

As he left the bridge to take command of the boarding party, Captain Velmar observed to one of his

Exhibit C
The Marlin Longline Wars

subordinates, with some dark amusement perhaps, that the confusion of tongues on the VHF had turned to stony silence. Once onboard the *Copemapro V*, however, he would soon discover that there were many other confusions to be dealt with.

For starters, the ship was crewed illegally. As a vessel of Mexican registry, she *should* have had a Mexican crew. She didn't. There were 18 Japanese "technical advisers" aboard and only three nationals.

There was also some confusion as to the actual identity of the vessel. What was her real name after all? Japanese letters were bleeding through the fresh paint just beneath the Spanish name on her hull.

And the ship's log was a sham. There were seven Mexican crewmembers listed, but only three could be accounted for. The billfish onboard were technically legal but their numbers had been grossly underestimated. Captain Velmar would later testify that in addition to all the dorado, tuna, and snook, there were more than 126 tons of marlin and sailfish in the holds. The log showed a *total* catch of only 45.5 tons of fish.

But it was even deeper within the refrigerated innards of the longliner that Captain Velmar found the most flagrant violation of his country's laws. Beneath a cosmetic layer of legal fish, he found an estimated two tons of illegal shark fins. According to a subsequent government report (obtained by PMY from Mexico City), the captain of the longliner later admitted that approximately 2,500 sharks, mostly young threshers, had been killed for their fins alone. After the animals were dismembered, the captain said, their bodies were simply dumped back into the sea.

Notice the black spray paint on the bow in this photograph taken by NOAA. Asian longline and driftnet boats obscure identifying markings to prevent observers from tracking their movements and estimating how many fish they may actually be taking from the sea.

The Trouble With Paradise

Ever since the protein-hungry Asians had come to fish Mexico's coasts for marlin, there had been problems. After a while, many of the people of Cabo and its environs began pressuring their government to control or stop the longlining "pirates" from across the sea. Someday, all the marlin would be gone, they argued, and so would the sportfishermen and the tourists. Cabo would die.

Letters, telephone calls, and petitions to Mexico City did little good. The Department of Fisheries maintained that stocks of marlin were plentiful, hardly in danger of depletion. There was enough to go around, the bureaucrats said, and permits to longline were issued freely.

Sportsmen and hoteliers responded to this liberality with allegations that commercial fishermen regularly violated permit restrictions by working within "protected zones." It was said that local authorities did not enforce laws made to conserve the marlin fishery. Charges of bribery and corruption were leveled at officials and, many people agreed, there were fewer and fewer marlin to be caught in the cobalt blue coastal waters near Cabo. The situation might have gone on unchanged for many more years, except for one thing. A new government came triumphantly to the nation's capital.

What Happened Next

Once the boarding party had returned to the gunboat, Captain Velmar decided to contact his superiors, via radio, about the irregularities he'd observed aboard the longliner. Acting upon their advice, he placed the *Copemapro V* under the direction of the Undersecretary of Fisheries in La Paz and ordered her to proceed to Cabo San Lucas where a more thorough investigation was to take place.

What happened next was perhaps predictable, given the way things of this sort had often been handled in the past. After several hours of biding her time in disgrace at the dock, the longliner was set free. All charges were dropped and the ship discretely left Cabo to resume operations at sea.

Public outrage was immediate. The story of the *Copemapro V* soon found its way to influential ears in Mexico City and in a very short time the new government there began its own nononsense investigation of the incident. On March 28, just a little over a month after the arrest of the *Copemapro V*, a communique from the office of the newly appointed *Secretaria de Pesca* in the capital was read publicly in Cabo.

The communique stated that three highly-placed officials in the Department of Fisheries—those who were immediately responsible for releasing the longliner—had been fired. Conceivably, more heads would roll in the future. The communique also stated that the *Copemapro V* had been charged with 1) making false log entries, 2) catching a species (sharks) not authorized by fishing permit, 3) destruction of marine animals, 4) fishing in "protected zones," and 5) maintaining a crew predominantly composed of non-Mexicans. The Copemapro fishing syndicate was fined 195,000,000 pesos (about $85,000) and its fishing permits were revoked.

Beyond Mexico

There's more to the *Copemapro V* story. But before getting to it, let's take a brief look at some other places in the world and how they too are threatened by industrialized fishing techniques. What's going on in Cabo is only part of a much larger problem. In the Gulf of Mexico, a population explosion is taking place. The American longline fleet there is expanding so rapidly that yellowfin tuna are being harvested like never before.

Exhibit C
The Marlin Longline Wars

Back in 1981, approximately 13 tons of yellowfin were taken from the Gulf during the year, according to statistics collected by the National Coalition For Marine Conservation. In 1988, less than a decade later, some *8,650 tons* were harvested and either shipped to Japan or sold fresh in the States. (See graph, next page.)

There is a direct relationship between this exponential growth and the dangerous depletion of billfish stocks. The reason for the connection is twofold. First, tuna longliners are taking *incredible* numbers of fish by comparison with older, more traditional methods. Second, longliners *can't* target a particular species, due to the nature of their fishing technique. They are constrained to catch and kill indiscriminately. So, the more tuna they catch, the more billfish they kill inadvertently.

In the past few years, the National Marine Fisheries Service has placed observers aboard tuna boats in the Gulf. These observers have determined that on average, one billfish is hooked per longline set. Most vessels make about 100 sets per year and, it's *conservatively* estimated, there are something like 250 longline vessels in the fishery today.

So how do these figures add up? Last year, during the hunt for tuna in the Gulf, it's safe to say that at least 25,000 billfish were hooked on longlines. Observer data shows that about one-half of them were dead when they were returned to the sea.

The Atlantic Swordfish

The effects of longlining for tuna in the Gulf are relatively insignificant when compared with the effects of mega-fishing on the high seas by giant Asian and American corporations. Oddly enough, the roles that these two, seemingly disparate commercial forces play in the ecology of the world's oceans today is, in some ways, rather complementary. To explain why, let's briefly examine how lawmakers in Washington presently look at tuna fishing.

First of all, the United States does not regulate the harvest of tuna in its own waters, although most other modern countries do. We claim that a *migratory* species cannot be realistically controlled by enforcing unilateral regulations. The tuna is a citizen of the world, the legislators say, and is thus outside the purview of one nation.

Congress officially came to this conclusion back in 1976, when it passed the Magnuson Act, a piece of legislation which was, prior to its enactment, subject to an intense lobbying effort by the big American tuna companies. Essentially, the lobbyists argued that restricting foreign tuna fishermen in American waters would only prompt other countries to do likewise along their own coasts. This, they said, would make life pretty difficult for American tuna boats working overseas. Don't regulate anybody, the lobbyists concluded, and everybody's happy.

Congress bought the argument, although many other countries at the time were already starting to regulate tuna anyway. Today, the "tuna exemption" is being used to subvert the conservation of other kinds of fish and marine life in American waters. Longline and driftnet fishermen, both foreign and domestic, catch otherwise protected species as a by-catch and excuse themselves by citing the Magnuson Act. We have a right to fish for tuna without restriction, they say. We can't be held responsible for killing other kinds of fish (and marine life) that get in the way.

A couple of years ago, the enormity of the problem was made painfully clear to conservationists on the Eastern seaboard. At the time, a plan was being developed to conserve the Atlantic swordfish, a species much in need of protection. According to that plan, swordfishing would be suspended during certain periods of the year. Since swordfish are primarily night feeders, longlining for tuna was to be restricted to daylight hours during these periods in an attempt to reduce the by-catch.

In opposition to the plan, the Japan Tuna Association filed suit against the U.S. government, claiming that the rights of Japanese longliners to work in American waters, as guaranteed in the Magnuson Act, would be violated by proposed regulations. The upshot? The Japanese won in the American courts and the plan to conserve swordfish never got off the drawing board.

This photograph was taken by an Earth Trust expedition member aboard a driftnet boat. Besides swordfish, these ships (Japanese, Korean, and Taiwanese) also catch and kill marlin and marine mammals, as well as massive quantities of tuna.

Driftnets In The Pacific

In a recent and lengthy report to Congress regarding the use of driftnets, NOAA took a stand which is particularly vehement given the organization's rather conservative nature: "Concerns that high seas driftnet fisheries are killing large numbers of marine resources of interest to the United States, and that these incidental takes could be affecting populations adversely *are* justified."

There are several major driftnet fisheries in the north Pacific, run by the Japanese, Koreans, and Taiwanese, targeting squid, salmon, billfish, and tuna. The Japanese have huge automated boats, and their nets can cover a slice of ocean some 40 miles long and 40-feet deep. Names and other identifying marks on the vessels are often spray-painted out, making it difficult for researchers to keep track of a particular ship's movements and the number of times she may return to the mothership to offload. A recent agreement between the United States and Japan which calls for the monitoring of driftnetting activities in the North Pacific will perhaps now provide some data where previously there was none. But, it will hardly improve the situation much more than that. The agreement includes *no actual restrictions* on the use of driftnets and longlines.

Earthtrust, a Honolulu-based international conservation organization, conducted an expedition to observe driftnet operations

Exhibit C
The Marlin Longline Wars

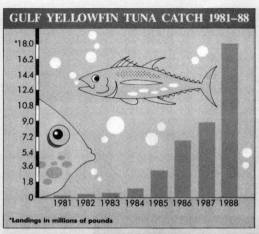

GULF YELLOWFIN TUNA CATCH 1981–88

*18.0
16.2
14.4
12.6
10.8
9.0
7.2
5.4
3.6
1.8
0

1981 1982 1983 1984 1985 1986 1987 1988

*Landings in millions of pounds

in the North Pacific last year. Earthtrust members returned with photographic documentation (some included in this article) which graphically shows the impact of driftnetting on billfish, marine mammals, and other sea life. The images, which have been featured in international publications, have helped focus attention on the problem and have sparked interest in a world-wide pelagic driftnetting ban.

Expedition member Sam LaBudde, a former U.S. government biologist, was able to spend some time aboard a working driftnet boat. One night's fishing, LaBudde reports, usually produces from four to five swordfish or blue marlin, six to eight sharks, one or two gooney birds, and "massive quantities" of tuna. Many billfish, he adds, are too large to be processed by the machinery onboard and are simply left to float away dead.

Back To Cabo

At presstime, the *Copemapro V* had disappeared like a ghost from the coastal waters of the Baja, along with the rest of her longlining sisterships. More irregular-

TURNING POINT

"That's *illegal*," said Bob Hayes, a Washington-based attorney who should know. Since it took effect last fall, Hayes has been working on the defense of the Atlantic Billfish Plan—a set of regulations prohibiting the *sale* of marlin and kindred species taken from Atlantic waters.

The plan is seen by sportsmen and conservationists as a step toward halting the depletion of the marlin fishery on the Atlantic and Gulf coasts.

Hayes, who's currently retained by the Coastal Conservation Association, is defending the plan from the efforts of the commercial fishing lobby. The commercial lobby charges that the plan will complicate the lives of commercial fishermen and contribute little to actual conservancy. They are suing to have the plan repealed.

Hayes' pronouncement came up during a telephone conversation we had in connection with this story. I was telling him about a commercial fisherman I'd met on Memorial Day. It seems the fisherman—who works out of an East Coast port—has been selling the marlin he takes as by-catch on his tuna longliner. The marlin were dead, the fisherman told me, and didn't bring

much of a price. He'd even thrown back those that were still green when he brought them up. He didn't question the legality—or *il*legality—of what he'd done. In fact, it all seemed perfectly logical to him. "At least I don't have them stuffed and hang 'em in my den," he grunted.

Having become something of an authority on the Billfish Plan myself—I'd been researching the subject for weeks—I was aware that selling billfish harvested from the Atlantic was prohibited.

I ran it by Hayes and talked with the people at the National Fisheries Institute (NFI)—Hayes' opponents in court. What I found was that the first part of the plan prohibits the sale of billfish in order to prevent development of a commercial market.

Other regulations in the plan restrict the type of gear that can be used and the size of billfish (with the exception of swordfish) that *recreational* fisherman can possess. Another regulation states that commercial fishermen *cannot* possess Atlantic billfish. They *must* throw back any billfish that they catch, dead or alive.

Billfish can only be caught (and

kept if they exceed a minimum size) by sportfishermen with rod and reel. Thus, the overall aim is to allow the highest number of fish to survive and to eliminate the incentive for commercial exploitation of the billfish resource.

What does all this *really* mean to the people involved? From the commercial fishermen's standpoint, it means more rules and regulations to contend with, the possibility of having to pay fines, a forced change in their fishing methods, and the closure of an entire fishery. Conservationists and sportfishermen, on the other hand, feel that finally a fishery is being conserved *before* it's been irreparably damaged. Additionally, they're pleased to get exclusive rights to the billfish resource.

Spokespersons on either side of the issue acknowledge that conservation *is* important. The debate, however, centers on exactly what steps must be taken to conserve the billfishery. "No *one* group stands for conservation," said Eldon Greenberg, attorney for the NFI. "If the fisheries are wiped out, there won't be any jobs for commercial fishermen. Talking about who's for or against conservation won't get you

Exhibit C
The Marlin Longline Wars

ities, it seems, were discovered and authorities were now "on the lookout" for the entire Copemapro fleet. According to our sources, investigators found that the vessels were legally registered in Mexico *and* in Japan. They were working under two flags—a violation of Mexican law.

Some people said that the welcome absence of the big, rust-streaked vessels signaled the end of the Marlin Wars in Cabo. Some said no...wait. At any rate, everyone presumed that the ships were now fishing beyond the 200-mile limit, in the safety of international waters. It was also presumed that the ships were drawing

supplies from Japanese support vessels out there and quietly slipping into ports in California to take on fuel when necessary. Copemapro fishing syndicate owner Mario Comperan declined to comment when contacted by PMY in Ensenada.

Luis Bulnes, spokesman for the hotel owners' association in Cabo, was not so tight-lipped. "You know I am what you call the old soldier of the Marlin Wars," he said. "And I know that in all the world, they have had the same billfish problem with the Japanese that we have here. In Australia and in Spain and in many other countries—I hear of the same

problem. You in the United States—you have the problems with the Japanese and others, like the Koreans and the Taiwanese.

"But I think the *real* predator is the *system* of fishing. It turns the ocean into a desert. These longliners and the gillnetters are a tough bunch—they kill the whales and the dolphins and the birds, too. Not only the billfish. The Asians turn their own coasts into a wasteland. If they are going to fish this way in the world, all the countries must get together and regulate them. Or else there will be nothing left for any of us."   □

TURNING POINT (CONT'D)

anywhere, you have to look at the facts."

The facts, according to the NFI's argument, are that there is no conservational intent behind the regulations. "How can the plan be conservation-oriented when sportfishermen can take as many marlin as they please," asserted Dick Gutting, vp of government relations for the NFI. "A commercial guy doesn't go out there looking for billfish. But if he does get some as by-catch, why shouldn't he be allowed to sell them?"

Sportsmen argue that if marlin can be sold without restriction, the number of billfish killed "incidentally" will rise in direct proportion to the price paid for them. Mike Leech, executive director for the International Game Fish Association: "Without the Billfish Plan, there may not be a direct commercial harvest of billfish. But...we suspect that fewer and fewer fish will be released (dead or alive) as their market value increases.

"Right now there is a market developing for these fish. If they become more salable, *no one* will be a recreational fisherman anymore. Everyone will be selling them."

The premise behind the plan is that if there's no demand for billfish, and hence no market, more fish will manage to survive. The effects of the plan are even more significant given the trend toward minimum-weight tournaments and tag-and-release fishing among sportsmen. "I think most sportfishermen realize that these fish are worth far more as a recreational resource and tourist attraction than they would be if they were sold individ-

ually," says Steve Sloan, an avid sportfisherman.

Economics

The economic impact of the plan is also a topic of hot debate. The commercial interests are concerned that the regulations will hamper their efforts to make a living. Many American fishermen want to see *international* laws enforced, so that restrictions on one nation do not indirectly benefit another. Still, sportsmen and conservationists maintain that the commercial point of view is shortsighted.

While the NFI claims it has no interest in developing a market for billfish, it *is* a distinct possibility. One conservationist suggested that if long-lining for tuna continues at the current rate, "the tuna fishery may get so small that *tuna* will become the by-catch of *billfish*. If that should ever happen, of course the billfish will be sold. That's business survival. Longliners will sell whatever they catch."

But it's not only commercial and sportfishermen who have a lot at stake in all of this. The billfishery means as much or more to many other people who may never even see a marlin. Citizens of fishing towns, boatbuilders and yards, gear manufacturers, and marina, restaurant, and hotel owners will all suffer dramatic losses if conservation efforts fail.

What About Japan And Korea?

As the battle over the Atlantic Billfish Plan goes on, the fact remains that the oceans are being fished more intensely than ever before. Current prac-

tices (the use of longlines and driftnets) constitute the most destructive fishing technology ever devised by man.

This case involves regulations which have been on the books for only a very short period of time. And it *is* a very important one because whichever way the court rules on the Atlantic Billfish Plan, the decision will be a cornerstone for future policy-making.

Meanwhile, it is suprising to note that everyone, whether conservationist or commercial fisherman, shares almost the same concern. The NFI wants to see regulations that "work." By that they mean management of the world's fisheries at an *international* level. They claim that they're not opposed to regulation.

But the Atlantic Billfish Plan, they argue, makes one country's conservation another's windfall. Listen to the longliner I met on Memorial Day: "Americans can conserve all the fish they want, Mike, but the boats from Japan and Taiwan and Korea and every place else will just take 'em all anyway. Then how am I supposed to feed my family? This thing's gotta be all or nothing."

Sportfishermen (and conservationists) essentially agree. However, they are convinced that to stop the worldwide depletion of the big billfisheries, someone, somewhere, must take the initiative.

And Bob Hayes believes he is working for people who are doing exactly that. During my last conversation with him, he summed it up: "The Atlantic Billfish Plan is the first step in the right direction. We *have* to start somewhere."   —Mike Micciche

Exhibit C
The Marlin Longline Wars

New England's Fishing Newspaper for over 20 years

Commercial Fisheries News

| une 1994 | A Compass Publication | Volume 21 Number 10 |

COMMERCIAL FISHERIES NEWS • JUNE 1994 • 21B

FRBs: **Money for groundfish rebuilding**

In a recent publication about the
ortheast fishery problems called
eyond Denial" by Charles H. Collins,
following facts were presented.
From 1976 to 1983, New England
hing effort doubled and stocks that had
en heavily fished by foreign fleets
:ame overfished in a few years.

GUEST COLUMN

From 1977 to 1983, New England
undfish populations declined 65%.
From 1977 to 1983, the number of
.e!   'll fisheries increased from
.. , but then declined to 1,334
.. , he number of fishing trips for
vessels increased 47% from 1977 to
(3. The total number of days fished by
:r trawlers increased 73% from 1973 to
(3.
'In 1993 groundfish landings
reased by 30%. Landings of cod,
Idock, and winter flounder in southern
nagement zones reached their lowest
:ls in history."
All of the above have contributed to
disaster we are now facing regarding
fishery stocks of the Northeast and, I

might add, throughout the eastern
seaboard. So far, we have attempted to
solve the problem by focusing on the
following, which is part of Sen. John
Kerry's fisheries recovery act:
● Developing markets for underutilized
species.
● Alternative fishing opportunities.
● Provide technical support and
assistance to fishermen and processors to
improve value-added fisheries and make
fishing for underutilized species
worthwhile.
● Developing methods to create
economic opportunities by processing fish
waste. (I assume this means bycatch).
● Aquaculture and hatchery programs to
restore depleted stocks.
The first three of the above will lead to
overfishing in "underutilized species" in
very short order, as those boats that have
been laid up seek income. Point four,
bycatch, is very important and must be
solved with specialized gear. It is point
five, aquaculture, which I believe holds
the most promise.
I have been invited by the Japanese
government and have visited their science
labs in Ishigaki Island and Kushimoto,
Japan, primarily to learn about their tuna
bio-mass science. During the visits, I was
shown other aquaculture projects, such as
butterfish rearing, grouper aquaculture,
and a fluke hatchery, where 10% of the
catch of commercial fishermen along a

100-mile coastline was attributed to the
two circular pens full of spawning fluke.
(This catch was in the millions of
pounds).
To restore our fisheries we need
conservation, and even denial
(moratoriums) plus + plus + plus + plus
aquaculture to put back species and
restore the stocks. How do we do all of
this at the same time?
We do it through the issuance of fish
recovery bonds (FRBs). In order to issue
(sell) a bond, you need a stream of income
that can be capitalized. This income
(bond payments) would come from a tax
on all items that go to sea, including:
anchors, fuel, line, rods, reels, lures, nets,

ice, hooks, winches, and vessels. That
income would be capitalized for 20 years
into an FRB.
For example: If the tax produced $50
million per annum for 20 years, a bond
amount of $600 million could be raised
and the payments on these bonds would
retire it at 5% interest in 20 years.
Why use a bond? The most logical
answer is that the income can be
capitalized and a great deal of money can
be raised. I believe that a great deal of
money is necessary to solve our fishery
problems.
Second, the bonds are a recognized
form of investment for pension funds,

See FISH BONDS, page 27B

COMMERCIAL FISHERIES NEWS • JUNE 1994 • 27B

Fish bonds *Continued from page 21B*

insurance companies, banks, and
trust companies. The bonds will
make it possible for state and
local institutions to invest in the
local economies to make sure of
the economic recovery. The
bonds, once sold, would have a
liquidity to the investor.
Third, the Band-Aid idea of
grants or disaster aid, even in the
amount of the $30 million
recently announced, will, I
predict, produce very little in the
way of direct results. We will
have more studies, more reports,
and little action. I see no other
way than FRBs to raise the
amount of money that is
necessary to do the job.
It is my proposal that the $600
million be put into a trust fund
and be administered in the
following fashion:
● Fifty percent ($300 million)
to the buyout or the
subsidization of mortgage
payments on vessels and
fishermen that have to sit out the
recovery of the fishery.
● Fifty percent ($300 million)
to a serious aquaculture
program to restore the decimated
stocks. The only reports paid for
in the program will be science
reports about the aquaculture
programs.
A board of directors with

National Marine Fisheries
Service (NMFS) should be
created to administer these trust
funds to monitor and review the
payments for both the fishermen
and the aquaculture programs.
The equation is simple:
Discipline + trust (fund)
payments = fisheries recovery.
This outline is a broad view
of FRBs. I am not politically
naive to think that lots of
massaging will take place.
However, without the
fundamental tool — money and
lots of it — we are in line for
more of the same painful finger-
pointing, wrenching meetings
that go nowhere, and worst of all,
the total collapse of our
groundfisheries. This collapse
will lead to further reductions of
all species along our coastal and
offshore waters.
The tool is money. As
Churchill said, "Give us the tools
and we will finish the job."
Stephen Sloan

*Stephen Sloan, chairman of
the Confederation of the
Associations of Atlantic Charter
Boats and Captains, first
introduced the idea of Fish
Recovery Bonds at a Maine
Fishermen's Forum tuna
seminar in March.*

**Mail form to Commercial Fisheries News, PO Box 37,
Stonington, ME 04681 • Fax 207/367-2490 • Call 800/989-5253**

Exhibit D
Description of Fish Recovery Bonds

Ocean Wildlife Campaign

Ocean Wildlife Campaign
1901 Pennsylvania Avenue. NW
Suite 1100 • Washington, DC 20006
202-861-2242 • Fax 202-861-4290

For Immediate Release   Contact:
November 24, 1998   David Wilmot (202) 861-2242

INTERNATIONAL MEETING BAD NEWS FOR ATLANTIC FISH:
Quota Increase For Depleted Tuna Cloaked As Recovery Plan;
U.S. Fails To Secure Needed Management Measures For
Declining Swordfish

National
Audubon
Society

The International Commission for the Conservation of Atlantic
Tunas (ICCAT) has once again failed to take appropriate action to
help severely depleted populations of bluefin tuna and swordfish
recover. While acknowledging that Western Atlantic bluefin tuna
are severely overfished, the Commission adopted a fundamentally
flawed "recovery plan", that cut by half the management target
that had been used for the past 23 years.

After abandoning management goals that had been its reference
point for over two decades, the Commission *increased* catch quotas
of the three nations fishing for Western Atlantic bluefin tuna
(Japan, the U.S., and Canada) despite the fish's severely
depressed status and declines of more than 80% since the mid-
1970s. "The consequences of this so-called recovery plan could be
disastrous. There can be no justification for a quota increase in
the initial stages of a rebuilding plan." said David Wilmot,
Director of the Ocean Wildlife Campaign and member of the U.S.
government delegation to ICCAT. "This just goes to show that you

WILDLIFE
CONSERVATION
SOCIETY

can achieve anything if you set your goals low enough," said Dr.
Carl Safina, Vice President for Marine Conservation at National
Audubon. "By halving the rebuilding target for these severely
depleted fish, ICCAT has declared that simply crossing the 50
yard-line is now a touch down."

WWF

Because fishers commonly get $10,000 (US) per bluefin for the
high-end sushi market in Japan (one fish was auctioned in Tokyo
for a record $83,000), commercial fishing interests exerted
tremendous pressure on the Commission to increase the catch quota.
Consequently, ICCAT rejected heeds for caution and chose to act on
an unrealistic rebuilding scenario. ICCAT decided to ignore
warnings from their own scientific committee that other valid
recovery scenarios projected Western Atlantic bluefin populations
could plummet within roughly 10 years unless catches were cut
significantly. Even the most optimistic scenario called for a
reduction in the quota, if the commission wanted to be "reasonably
sure" of maintaining the status quo.

In addition, ICCAT dealt North Atlantic swordfish a blow through
inaction. The U.S. recently issued a draft domestic recovery plan
for North Atlantic swordfish that it feels is dependent upon
international actions for full implementation. However, the U.S.
failed to secure the guidance it considers necessary from ICCAT to
fully implement the domestic recovery plan for swordfish. "The
failure of the U.S. to pursue needed measures

Printed on recycled paper with 50% post-consumer waste using soy-based inks
--MORE--

Exhibit E
Ocean Wildlife Campaign News Release

at the ICCAT meeting raises serious questions about its commitment to follow through on its forward-looking proposal to rebuild swordfish," said Lisa Speer, of the Natural Resources Defense Council and Co-founder of the Give Swordfish a Break Campaign. While the U.S. did secure a resolution to develop a rebuilding plan at ICCAT next year, a critical opportunity was missed to begin the swordfish rebuilding process. Swordfish populations remain at all-time lows and are still declining.

Most disturbing was the failure of the U.S. to seek an agreement by ICCAT to deduct dead discards of swordfish from country quotas. Since 1992, U.S. fishermen have killed and discarded an average of 30,000 - 40,000 undersize swordfish each year. In 1997, dead discards of juvenile fish amounted to 455 metric tons over and above the U.S. catch allowance. "Without needed measures to reduce this obscenely wasteful bycatch, such as closing known nursery areas and modifying fishing practices, failure to count this additional source of mortality against the allowable catch will further drag down the recovery of swordfish," says Ken Hinman, president of the National Coalition for Marine Conservation.

Progress was made in the area of compliance with ICCAT regulations. However, even full compliance with the current inadequate management measures will not achieve rebuilding of depleted fisheries to healthy and productive levels. For example, ICCAT has established--for 1999 and 2000--a total allowable catch for Eastern Atlantic bluefin tuna that its own scientific Committee has determined is not sustainable.

In addition, after years of battling to open up the ICCAT process to observation by non-governmental organizations (NGOs), ICCAT revised its guidelines for granting observer status. These changes will allow conservation organizations and fishing groups to participate in on-going efforts to force ICCAT to take its stewardship responsibilities more seriously.

The United States' failure to indicate support for a real rebuilding scenario for Western Atlantic bluefin tuna in its draft fisheries management plan, prior to attending the ICCAT meeting, contributed to ICCAT's acceptance of a flawed recovery plan with reduced recovery goals. Further, ICCAT's failure to adopt important management provisions for North Atlantic swordfish has made the U.S. draft swordfish recovery plan a hollow shell. As a result, the only way the U.S. National Marine Fisheries Service can ensure that swordfish are rebuilt to healthy and productive levels within ten years, as required under U.S. law, is to immediately enact the most stringent management measures allowed under the Magnuson-Stevens Act.

The OWC is a coalition of National Audubon Society, National Coalition for Marine Conservation, Natural Resources Defense Council, Wildlife Conservation Society, and World Wildlife Fund. It was created to tackle the complex challenge of conserving and restoring large ocean fishes, including sharks, tunas, and billfishes. The OWC receives primary funding from The Pew Charitable Trusts.

Exhibit E
Ocean Wildlife Campaign News Release

Nov/.1997
Doc #71

RECOMMENDATION BY ICCAT REGARDING BILLFISH

RECOGNIZING that the objective of ICCAT is to maintain populations of tunas and tuna-like species, including Atlantic blue marlin and Atlantic white marlin, at levels that will produce maximum sustainable yield (MSY) for food and other purposes;

EXPRESSING CONCERN that the SCRS has estimated that Atlantic blue marlin biomass is at 24% of the MSY level, and that Atlantic white marlin biomass is at 23% of the MSY level;

NOTING that the current stock assessment is derived in part from problematic data as indicated in the SCRS report;

CAREFULLY REVIEWING projections for blue marlin and white marlin indicating that reductions in fishing mortality are necessary to avoid further declines in the stocks and to begin rebuilding these stocks;

RECOGNIZING that rebuilding blue marlin and white marlin stocks will be beneficial to all parties fishing on these stocks, as they are a source of food and recreational activity for many countries;

THE INTERNATIONAL COMMISSION FOR THE CONSERVATION OF ATLANTIC TUNAS (ICCAT) RECOMMENDS THAT:

All Contracting and Non-contracting Parties, Entities, and Fishing Entities

(1)   Reduce, starting in 1998, blue marlin and white marlin landings by at least 25% for each species from 1996 landings, such reduction to be accomplished by the end of 1999.

(2)   Promote the voluntary release of live blue marlin and white marlin.

(3)   Advise ICCAT annually of measures in place or to be taken that reduce landings or fishing effort in the commercial and recreational fisheries that interact with blue marlin and white marlin.

(4)   Provide all base data requested by the SCRS to improve stock assessment and work to improve current monitoring, data collection and reporting procedures in all their fisheries. In 1999, the SCRS shall conduct blue marlin and white marlin stock assessments and, at the 1999 Commission meeting, the Commission shall review the stock assessment and recommend appropriate management measures, if necessary.

(5)   The provisions of Section 1 shall not apply to small-scale artisanal fisheries, i.e. those small-scale fisheries for subsistence purposes, including sale to local markets.

The ICCAT Secretariat shall inform all Non-Contracting Parties, Entities, and Fishing Entities of this recommendation and encourage them to cooperate with these measures.

United States Department of State

Exhibit F
Recommendation by ICCAT Regarding Billfish,
Document 71

Attachment 3

**Report of the Bigeye, Albacore, Yellowfin,
and Skipjack (BAYS) Tunas Working Group**

Advisory Committee to the U.S. Section to ICCAT
2001 Species Working Group Workshop

Holiday Inn, Silver Spring
April 9-10, 2001

Dr. John Mark Dean, Convener
Kim Blankenbeker, Rapporteur

I.   **Recommendations for BAYS Tunas**

A.   Data

1)   NMFS should meet with its state counterparts and the ACCSP to develop improved recreational and commercial landings data and develop new mechanisms to collect data in all regions throughout the year. NMFS should develop a certification program for states and/or organizations (such as universities) within those states to delegate data collection efforts. See attached statement.

2)   Given the insufficiency of the U.S. landings data reported to ICCAT, NMFS should report recreational BAYS landings as provisional, and these data should not be used for future allocation purposes.

B.   Research

1)   The United States should conduct life history studies on BAYS tunas, including age and growth, reproduction, stock structure, and essential fish habitat studies. NMFS is encouraged to have as much of this research as possible conducted by third parties.

2)   NMFS management should prepare a report in table form for the Spring 2002 Advisory Committee meetings, and annually thereafter, that consists of a detailed listing of research projects on HMS species. Specifically, the table should be composed of the following elements: Project title, investigator, affiliation, level of funding or budget allocation, man-years of effort, source of funding, time frame, and type of deliverable. (The table should list projects from all sources, including but not limited to: NMFS, SeaGrant, Marfin, S-K, Wallop Breaux, outside foundations (e.g., Packard Foundation, Pew Trust), states, etc.)

3)   Economic impact and economic benefit studies of recreational fisheries that target BAYS tunas should be conducted. An outline proposal of what the proposed economic impact and economic benefit studies should be provided to the Advisory Committee for discussion at the Spring 2002 Committee meeting.

C.   Management

6/20/02

Exhibit G
Report of BAYS Working Group

200

1) The United States should pursue international rebuilding programs for all over fished BAYS tuna stocks, in particular northern albacore and bigeye, considering their domestic importance. Given the nature of the U.S. fisheries for BAYS tunas, the United States should only be responsible for a proportional share of the burden associated with rebuilding.

2) The United States should seek to ensure that nations harvesting bigeye and yellowfin tunas comply with ICCAT's 1997 recommendation to improve compliance with minimum size recommendations and take steps to accelerate the implementation of measures to reduce the harvest of these tunas that are less than the minimum size, including implementing the provisions of the minimum size compliance measure.

3) Recognizing that the Gulf of Guinea is an important area for recruitment in the BAYS tunas fisheries, ICCAT should continue to direct efforts to investigate these fisheries and develop effective measures to reduce the mortality of sub-legal fish in this area, including continued analysis of such measures directed at rebuilding tuna stocks.

4) Consistent with number 3 above, SCRS should continue to identify spawning grounds and other areas of high concentration of juvenile BAYS tunas (along with other HMS) and recommend international time/area closures and/or sanctuary areas for the BAYS tunas and other HMS, where needed.

5) The United States should seek a measure that would reiterate the responsibility to provide basic catch data and establish penalties for non-reporting. Specifically, ICCAT could create a process analogous to the UII process that would identify those ICCAT members egregiously violating catch data reporting requirements.

6) The United States is encouraged to provide technical support/assistance to developing states to assist in the development of effective fishery management practices, including data collection and reporting, in order to facilitate the full participation of these countries in ICCAT and their adherence to ICCAT's conservation and management measures.

II. **Recommendations for Other Species**

A. Sharks

1) The United States should actively pursue that SCRS conduct stock assessments for blue, mako, and porbeagle sharks as soon as possible after the 2001 data preparatory meeting and no later than 2002.

2) The United States should vigorously pursue an agreement to eliminate the practice of retaining only the fins of sharks while discarding the carcasses. The provisions of the U.S. Fishery Management Plan for Atlantic Highly Migratory Species pertaining to the ratio of fins to carcasses landed should be presented to ICCAT.

3) In the ICCAT forum, the United States should encourage the live release and tagging, when possible, of juvenile sharks.

4) The United States should conduct life history studies on sharks, including age and growth, reproduction, stock structure, and essential fish habitat studies.

6/20/02

Exhibit G
Report of BAYS Working Group

58

NMFS is encouraged to have as much of this research as possible conducted by third parties.

B.   Wahoo

The United States should urge ICCAT to collect Task I data for wahoo, so that a database for this species can be established.

III.   **Other Issues**

A.   Monitoring

1)   NMFS should conduct a workshop to identify cooperative research with regard to at-sea data collection methods (e.g., observers, VMS, electronic logbooks) with a view to moving these issues into the ICCAT arena in the future.

2)   The BAYS Working Group supports Japan's large-scale monitoring initiative in principle, but notes a number of items, such as VMS and real time data reporting, could be applied more broadly in ICCAT fisheries than contemplated in the Japanese proposal. Additional analysis of these applications should be conducted by the United States. In addition, the United States should consult with Japan regarding the possible addition of new items, such as the development of an ICCAT flagging program which would clearly indicate those vessels that are permitted to fish in the Convention area and the addition of language on compliance and on technical assistance.

C.   Future of ICCAT

1)   The working group is generally supportive of the Commissioner's initiative concerning the future of ICCAT, and would like to stress its concern for the continued viability of the organization.

2)   The Working Group would like to make the following motion for consideration by the full Committee:

We, the members of the ICCAT Advisory Committee, strongly recommend that the Secretary of Commerce support the continuance of Rolland A. Schmitten as U.S. Federal Government Commissioner and head of the U.S. Delegation to ICCAT. There are many important issues now at critical stages of negotiation in ICCAT, including critical compliance issues. In fact, the viability of ICCAT itself is at stake. Therefore, the Committee believes a continuity of effort would be in the best interests of the Commercial, Recreational, and Environmental interests as well as those of the United States.

BAYS Tunas Working Group Members

Eleanor Bochenek (Technical Advisor)
Virdin Brown
John Mark Dean (Convener)
Bob Eakes (Absent - temporarily reassigned to Billfish)
Bob McAuliffe (Technical Advisor)
Ernie Panacek (Technical Advisor)
Ellen Pikitch
Greg Skomal (Technical Advisor)
Steve Sloan

6/20/02

Exhibit G
Report of BAYS Working Group

Randi Parks Thomas (absent)

<u>Other Attendees</u>

Kim Blankenbeker (NMFS, Rapporteur)
Nikki Brajevich (Department of State)
Jerry Scott (NMFS SEFSC)

59

Exhibit G

Report of BAYS Working Group

Attachment to BAYS Tunas 2001 Report

(Submitted by Steve Sloan to the 2001 BAYS Tunas Working Group)

After review of the 2001 BAYS tunas working group report, I wish to correct an inadvertent error. For the years 1994-2000 inclusive, and now 2001, the BAYS working group has had requests from recreational members to include language in its reports that objected to the amount of yellowfin tuna catch reported by NMFS to ICCAT. For the aforementioned years, the recreational community believes the true amount of recreational yellowfin catch is between 18,000 and 22,000 metric tons. NMFS has refused, despite many requests, to do the requisite study to determine the correct amount of yellowfin tuna recreational catch. A case in point was northern albacore where the United States was held to less than two percent of the TAC because of the previous ten-year average. The northern albacore TAC for the United States was determined to average 600 metric tons per annum. Therefore, if this system is used in the event TACs are established for yellowfin tuna, the U.S. recreational quota would be the fraction 5,000/165,000 metric tons instead of 18,000-22,000/165,000 metric tons, a difference of over 300 percent. Before NMFS disenfranchises the yellowfin tuna recreational fishery and fishermen, a correction of the current number is in order (which has been outlined and requested since 1994). A correct count of the recreational yellowfin tuna fishery is once again urgently requested. (Signed: Stephen Sloan)

Note: It was determined at November 2000 ICCAT meeting that the ten-year average recreational catch of northern albacore was 450 metric tons and commercially was 150 metric tons.

6/20/02

Exhibit G
Sloan Attachment to Report

Exhibit H

Brochure of Spanish Shipyard

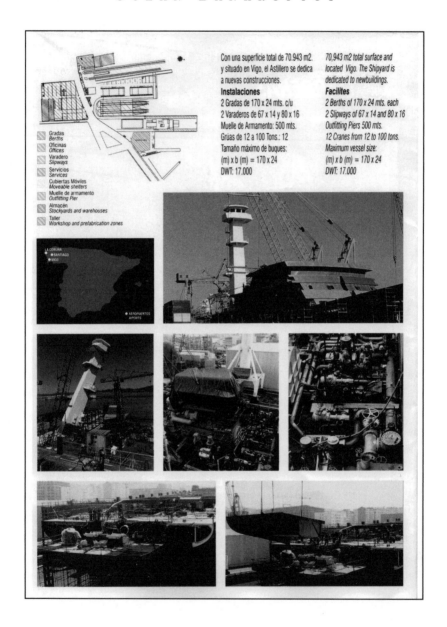

Exhibit H
Brochure of Spanish Shipyard

Exhibit H

Brochure of Spanish Shipyard

Exhibit H
Brochure of Spanish Shipyard

Exhibit H

Brochure of Spanish Shipyard

Exhibit H
Brochure of Spanish Shipyard

SCRS/00/95

SUMMARY OF POP-UP SATELLITE TAGGING OF GIANT BLUEFIN TUNA IN THE JOINT US-CANADIAN PROGRAM, GULF OF MAINE AND CANADIAN ATLANTIC*

by

Molly Lutcavage,[1] Richard Brill,[2] Julie Porter,[3] Paul Howey,[4] Edward Murray, Jr.,[5] Anthony Mendillo,[6] William Chaprales,[7] Michael Genovese,[8] and Ted Rollins[4]

SUMMARY

Since 1997, we deployed 58 single point and 21 light-sensing pop-up archival satellite tags (Microwave Telemetry, Inc., Columbia, MD) on giant bluefin tuna (178-266 cm SFL) in the western North Atlantic. The goals of our initial deployments were to test external tag attachments and the tags themselves, which evolved to include greater data logging capacity, additional sensors, and increased power. We deployed all of the tags on fish from New England and Canadian commercial or charter fishing vessels (harpoon, rod and reel, trap, and purse seine) using tag attachment techniques developed by the U.S. fishermen (Murray, Chaprales, Mendillo, and Genovese). Attachment periods ranged from 5 - 365 days, although the majority of tags detached from the fish over the presumed spawning period (April - July). Tag reporting success rates were 59% for single point tags and 79% (15 out of 19 due) for the archival tags. Three tags (shed from large fish in Canada) reported from land. Without exception, results from 1997-2000 tagging showed that all tagged fish were in the central Atlantic when their tags reported, and 30-58 % annually were in the eastern management area. We attribute our high tag reporting success rates to the experience of fishermen tagging partners, appropriate handling of the bluefin tuna during capture and tagging, careful tag placement, and proper storage and handling of the satellite tags. We now have data capable of depicting full migration paths and environmental associations (80 - 270 days) of ten fish (193-266 cm). Data successfully returned from the archival tags will generate geolocation estimates and errors associated with light-derived data. In 2000, we plan to deploy pop-up archival tags for 365-500 day attachments.

* Not to be cited without the permission of the authors.

---

[1] Edgerton Research Lab, New England Aquarium, Central Wharf, Boston, MA, 02110 [2] Honolulu Lab, SWFSC, NMFS, Honolulu, HI [3] Fisheries and Oceans Canada, Biological Station, 531 Brandy Cove Road, St. Andrews, NB, E5B 2L9 [4] Telemetry 2000, Inc., 10280 Old Columbia Rd., Columbia, MD, 21046 [5] 8101 Nashua Drive, Palm Beach Gardens, FL, 33418 [6] Kene Boats, Isla Mujeres, MX [7] PO Box 285 Marstons Mills, MA, 02648 [8] White Dove Inc., Cape May Courthouse, NJ 08210

Exhibit I
SCRS Report on Tuna Tagging

1. INTRODUCTION

Beginning in 1997, in a collaboration between researchers and fishermen, we conducted pop-up satellite tagging of spawning size class Atlantic bluefin tuna (*Thunnus thynnus*) in the NW Atlantic. These tags, developed by Microwave Telemetry, Inc (Columbia, MD), consist of a radio transmitter, environmental sensors, and data logger designed to jettison from the fish after a predetermined release date (Block *et al.*, 1998; Lutcavage *et al.*, 1998). Because the fish does not have to be caught and tag returned to retrieve the data, the pop-up satellite tag is the first fishery-independent tool to deliver information on a fish's movement over large spatial and temporal scales. Our initial goals were to develop and test long-term external attachment methods developed by the U.S. fishermen (Chaprales, Murray, Genovese, and Mendillo) and to deploy the new pop-up tags on giant bluefin released from New England commercial bluefin harpoon, rod and reel, and purse seine fisheries. Our long-term goals were to determine the movements and behavior of the bluefin tuna assemblages on the New England shelf in the summer and fall.

In 1998, our tagging program included releases from Canadian trap, rod and reel, and harpoon fisheries. We targeted adult fish comprising spawning size classes (>200 cm SFL) and programmed the majority of the tags to detach from the fish over their presumed spawning period (April - July). Results from our 1997-1998 deployments were consistent: all successfully released tags reported from the central Atlantic roughly between Bermuda and the Azores. Each year, about 30% of the tags reported from east of the 45° W. stock division line and none of the giant bluefin were in or near known spawning grounds (Gulf of Mexico or Mediterranean Sea (Lutcavage *et al.*, 1998; Lutcavage *et al.*, 1999a; 1999b).

In 1998 we also tested short-term (3-30 days) prototype pop-up archival tags (PTT-100, Microwave Telemetry, Inc., Columbia, MD), which have an expanded data logging capacity and a light sensor. After data processing, recorded ambient light levels allows determination of day length and local noon to estimate latitude and longitude (*e.g.*, Klimley *et al.*, 1994; Welch *et al.*, 1999). Combining desirable properties of conventional (fishery-dependent) implanted archival tags (Restrepo, 1996; Block *et al.*, 1998) with the fishery-independent pop-up tag, pop-up archival tags reveal fine to mesoscale information on position, temperature, and depth of the tagged fish. In composite, behavior of the tagged fish can then be reconstructed. In 1999, we deployed 21 pop-up archival tags and eight single point pop-up tags on spawning size class bluefin tuna for attachment periods of up to 365 days, and here report preliminary findings from all tags reporting by 1 September, 2000.

2. MATERIALS AND METHODS

In 1999, seven giant bluefin were tagged and released from a purse seiner and 22 were tagged after capture by rod and reel following methods developed by our tagging team (Lutcavage *et al.*, 1998; 1999). Six of the 29 fish were tagged in Canadian waters. Four of the pop-up archival tags were prototypes scheduled to release after 5-30 days (short term tags), and the remaining seventeen tags in late December, 1999, and in April, July, and September, 2000. Single point tag attachments (8 tags) were set for 225-286 days. All of the pop-up tags included a conventional fish identification tag (National Marine Fisheries Service) printed with tag reporting information and notification of reward.

Single point pop-up tags stored 61 mean values (1.0 - 4.8 day intervals) of ambient temperatures (± 0.2° C) measured once an hour. Archival tags recorded ambient light levels at 2 min intervals and temperature (± 0.2° C) once an hour. Longitude and latitude estimates were generated with proprietary software written by Microwave Telemetry, Inc. and have an associated error of about 1° longitude and 2° latitude, although more accurate geopositions may be calculated following additional data processing (Howey *et al.*, in preparation). Lengths and weights of fish were estimated by the captain and mate (who

Exhibit I
SCRS Report on Tuna Tagging

had 20-25 years of commercial bluefin fishing experience). In some cases weights were converted to lengths based on recent NMFS conversions tables (Dr. Steve Turner, SEFSC, Miami, FL).

3. RESULTS

Fifteen out of nineteen (79%) pop-up archival tags (that were due) reported their release positions and successfully returned archived data (Table 1). One of these tags was shed (or not successfully attached) from a fish released off northeastern Nova Scotia and reported from land, not far from the capture location, and four pop-up archival tags were never heard from. Only two out of eight single point pop-up tags (12%) reported their location and data. Estimated sizes of bluefin tuna bearing tags that successfully reported data on schedule were 190- 200cm (2 fish), >200-210 cm (2 fish), >210-220 cm (2 fish), >220-240 cm (3 fish), and >240-270 cm (3 fish).

Tag reporting positions of bluefin tuna released in 1999 (long term tags only, i.e. > 80 day deployments) are shown in comparison with results from previous field seasons (1997-1998) (Fig. 1). Seven out of 12 fish were found in the eastern management area, and none were located in either the Gulf of Mexico or the Mediterranean. With the exception of one fish (205) that reported in mid April from 21° W, all of the tagged fish were in the central Atlantic east of the Gulf Stream and west of the mid-Atlantic ridge.

A longitude time series versus Julian date for all reporting pop-up archival tags shows that all of the bluefin tuna traveled east within 30 days of their release (Fig. 2). With one exception (Fish 205), fish with tags reporting between April and 30 June (6 tags) eventually slowed their rate of eastward travel (upward curving slope). The two fish with tags reporting on 30 June were east of Bermuda and gave no indication of migration west back to the U.S. continental shelf. Two fish captured by purse seine and released with fish 212 and 216 migrated directly to the southeast along similar routes (Fig. 3), although one traveled farther south, reaching the Bahamas by early November. By mid November both fish headed to the northeast along the margins of the subtropical convergence, but were over 2,500 km apart on their tag reporting date (25 December).

Preliminary geoposition estimates for four fish whose tags jettisoned in April, 2000 (Fig. 4) show travel routes that are generally similar to the December 1999 reporting fish (fish 211, 215, 212 and 216 were released from the same school). All four migrated to the southeast but by mid-April, three were in the eastern management area. Estimated migration paths for all of the fish tagged with pop-up archival tags overlay single point tag reporting locations (Fig. 1). Geolocation estimates are in preparation for July releases, and light and temperature data from are still being transmitted from two tags that jettisoned from two fish on 1 September, 2000.

A summary of ambient temperatures recorded from fish with tags reporting in December 99 and in April 2000 is given in Table 2. In the New England region bluefin experienced ambient temperatures of 5-21° C, but by mid-April, in the central Atlantic their maximum daily temperatures (except for fish 205) were 24 - 28° C. Temperature time series from these fish became stable in April – July, and in general, varied less than 3-5° C. Fish 205 was in high latitudes from Jan-April, and in considerably cooler water (<14° C) than the others.

4. DISCUSSION

Based on hourly temperature records, the tag that was recovered on land in northeastern Nova Scotia was attached to the fish, but was shed after about ten days. In contrast, two tags previously

Exhibit I
SCRS Report on Tuna Tagging

recovered on land in Canada probably were shed immediately (Lutcavage *et al.*, 1999). We suspect that differences in data reporting rates between the U.S. and Canada may be partly attributable to a longer fishing season and different fishing techniques, and the fact that the U.S. fishermen have significantly more experience with tag and release than our Canadian participants. This is likely to be less of a concern as the Canadian team gains experience. Ten tags (4 pop-up archivals, 6 single point tags) were not detected. Possible causes of non-reporting include tag or antenna damage, dive depths surpassing the float's rated limit (1,000 psi), natural mortality, faulty tag placement, and/or fishing mortality and non-reporting of tags by fishing vessels. As in previous years, it is unlikely that mortality from capture and tagging activities is the main cause of non-reporting, since tags would have eventually released from the fish, returned to the surface, and reported a position.

In comparison with our 1997-98 results (80% and 43%, respectively), the low reporting rate of single point pop-up tags in 1999 (12%) was disappointing, but since only eight were deployed, we're unable to identify the most likely source of failure. The European Union's bluefin tagging program (DeMetrio *et al.*, SCRS/99/55) has a lower overall single point tag data reporting rate (<40%), but in some cases their tags transmitted data but the signal was insufficient to determine a position (Dr. Geoff Arnold, personal comm., CEFAS, Lowestoft, UK). This was not the case for any of our tags.

In general, the 79% success rate of the pop-up archival tags deployed for up to one year was equivalent or better than single point tags, suggesting that the technology is reliable and that external attachments can be extended for longer periods. As in previous years, we attribute our high tag reporting success rates to the experience of our fishermen partners, appropriate handling of the bluefin tuna during capture and tagging, careful tag placement, and proper storage and handling of the satellite tags.

Data from the first long term ($\geq$ 80 days) pop-up archival tags show that giant bluefin tuna left the New England feeding grounds within 10 to 30 days and traveled directly to low latitudes (20-35° N) at estimated speeds of 5 - 8 kts. Although our data show that these fish traveled south past North Carolina, they did not linger near the inshore winter fishery described by Block *et al.* (1998; SCRS/99/103). The majority of bluefin tuna released with implanted archival tags from the North Carolina fishery and recaptured in New England/Canada were reported to have spent most of their time in the "western" Atlantic (presumably, west of 45° W), with the exception of three Mediterranean recoveries (Block *et al.*, SCRS/99/103). These fish were primarily age class 7 – 8, whereas most of the fish that we tagged in New England/Canada were within age classes 9 - 10+. Whether there is a major link between large giants and the North Carolina assemblage remains to be seen, but most of the fish landed there are smaller than the New England/Canadian assemblage, and conventional tag returns do not suggest that this is the case.

Tagged fish had bimodal maximum daily temperature records first indicating their Gulf Stream crossings and then their arrival and residency in warmer waters of the central Atlantic. The pop-up archival tag did not have a depth sensor, but stable temperature time series suggests that fish in the central Atlantic region spent most of their time at shallow depth (< 150 m), similar to patterns described for giant bluefin on the New England shelf (Lutcavage *et al.*, 2000) and for North Carolina fish (Block *et al.* SCRS/99/103). From there, all six of the fish with tags reporting in December and April headed to the northeast along the subtropical convergence boundary to the central Atlantic. Environmental analyses have not been completed for fish with pop-up archival tags reporting in July and September, 2000.

Including this year's results, a total of 36 tags released from spawning size class bluefin tuna reported their positions (in late January through 1 September) from the central Atlantic. Of those, 15 reported in April through July, spanning the presumed spawning period in both the western Atlantic and Mediterranean Sea (Rivas, 1954; Richards, 1976). Block *et al.* (SCRS/99/103) reported that 20 pop-up

Exhibit I
SCRS Report on Tuna Tagging

tags jettisoned from bluefin tuna outside of the Gulf of Mexico from April to June, although not all of these fish were believed to be mature. So far, none of our pop-up satellite tagging results challenge the hypothesis that some Atlantic bluefin tuna may spawn in the central Atlantic, possibly along frontal areas north and east of Bermuda and west of the Azores (Lutcavage *et al.*, 1998; 1999). It seems very unlikely that all of the mature bluefin tuna that we tagged were on an alternate-year spawning cycle, as this has not been found to be the case in any other tuna species. Block *et al.* (SCRS/99/103) recently came to the conclusion that two implanted archival tags from two bluefin tuna tagged in North Carolina in 1997 and recovered in 1999 showed no residence in the Gulf of Mexico or Mediterranean during the presumed spawning season, although several others were captured in each of the known spawning area.

Several researchers have suggested that bluefin tuna spawn in summer months along the U.S. continental shelf near the Gulf Stream edge (Mather and Bartlett, 1962; Baglin, 1974; Mather *et al.*, 1995), and more recently, large ovaries have been sampled from commercial landings of New England giant bluefin (B. Chase, Mass. Div. Marine Fisheries, personal comm.). We have also collected enlarged ovaries and testes with running "milt" in Gulf of Maine bluefin sampled this season (J. Goldstein, Univ. of Mass., unpubl. data). Preliminary reproductive studies on Atlantic bluefin tuna suggest that they become reproductively active within a 2-3 week window (Dr. C.R.. Bridges, personal comm., Univ. of Duesseldorf, Germany), so it is worth noting that some of these fish may have recently left their spawning areas. Muscle and gonad tissue will be submitted for reproductive and gonadotropin analysis to confirm their status.

In a workshop held in Bermuda in May, 2000, participants reviewed historic and recent satellite tagging information on the biology of central Atlantic bluefin tuna and produced a consensus document and draft research agenda addressing existing gaps in knowledge (Lutcavage and Luckhurst, SCRS/00/125). As the management implications of the presence or lack of spawning there would be enormous, the Bermuda working group stressed the urgent need to determine the reproductive status and possible spawning activity of central Atlantic bluefin tuna. The group concluded that an oceanographic research cruise and long-line sampling would offer the most effective means of resolving this critical issue, and would also allow fish to be tagged and released with satellite tags in the central Atlantic.

The resolution of geolocation data returned from the present generation of light-sensing archival tags (included pop-up and implanted tags) is tag-specific and strongly affected both by extrinsic (*e.g.*, ambient light, fish depth, weather) and intrinsic factors (*e.g.*, sensor type, electronics, software)(Klimley *et al.*, 1994; Welch *et al.*, 2000). However, errors associated with geolocation estimates are small in relation to the large-scale movements of Atlantic bluefin tuna, and the geolocation data is clearly able to identify their migration paths. As many other investigators have proposed, Atlantic bluefin tuna have complex life histories and migration patterns, and researchers have yet to fully understand their reproductive habits. Results from satellite and archival tagging studies so far seem to confirm early workers' views (Tiews, 1963; Hamre, 1963; Mather *et al.*, 1995) that over time, bluefin tuna of different age classes and assemblages make dynamic shifts in migration routes and feeding grounds. How this complex historic picture can be integrated with recent biological and oceanographic information to define the nature of the Atlantic bluefin population and stock structure remains a tantalizing question. The synthesis of data returned from all electronic data recording tags from all areas of the Atlantic will help to complete an accurate view of bluefin biology and migration and support restoration of the population.

5. LITERATURE CITED

BAGLIN, R.E., Jr. 1982. Reproductive biology of western Atlantic bluefin tuna. Fish. Bull. 80(1): 121-134.

Exhibit I

SCRS Report on Tuna Tagging

BLOCK, B.A., H. Dewar, C. Farwell, and E.D. Prince. 1998. A new satellite technology for tracking the movements of the Atlantic bluefin tuna. Proc. Nat. Act. Sci. 95:9384-9389.

BLOCK, B.A., H. Dewar, S.B. Blackwell, T. Williams, E. Prince, C. Farwell, A. Boustany, and A. Seitz. 2000. Archival tagging of Atlantic bluefin tuna. . ICCAT, Coll. Vol. Sci. Pap. SCRS /99/103.

Demetrio, G., G. Arnold, J.L. Cort, J.M. De la Serna, C. Yannopoulos, P. Magalofonou, and G. Sylos Labini. Bluefin tuna tagging using "Pop-ups": First experiments in the Mediterranean and eastern Atlantic. ICCAT, Coll. Vol. Sci. Pap. SCRS/98/55.

HAMRE, J. 1963. Tuna tagging experiments in Norwegian waters. Proc. World Sci. Meeting on the Biology of Tunas. FAO, Rome.

KLIMLEY, A. P., E.D. Prince, R. Brill, and K. Holland. Archival tags 1994: present and future. NOAA Technical memorandum NMFS-SEFSC-357. September, 1994. US Dept. of Commerce. 30 pp.

LUTCAVAGE, M., R.W. Brill, J.M. Porter, G.B. Skomal, B. Chase, P.W. Howey, and E. Murray, Jr. Preliminary results from the joint US-Canadian pop-up satellite tagging of giant bluefin tuna in the Gulf of Maine and Canadian Atlantic Region, 1998-99. ICCAT, Coll. Vol. Sci. Pap. SCRS/99/104. (In press)

LUTCAVAGE, M.E, R.W. Brill, J.L. Goldstein, G.B. Skomal, B.C. Chase, and J. Tutein. 2000. Movements and behavior of adult North Atlantic bluefin tuna (*Thunnus thynnus*) in the northwest Atlantic determined using ultrasonic telemetry. Marine Biology. *In press*

LUTCAVAGE, M. and B. Luckhurst. Consensus Document: Workshop on the biology of bluefin tuna in the central Atlantic. SCRS/20/125 (submitted).

LUTCAVAGE, M., R. Brill, G. Skomal, B. Chase, and P. Howey. 1999a. Results of pop-up satellite tagging on spawning size class fish in the Gulf of Maine. Do North Atlantic bluefin tuna spawn in the Mid-Atlantic? Can. J. Fish. Aquat. Sci. 56:173-177.

LUTCAVAGE, M., R. Brill, G. Skomal, B. Chase, and P. Howey. 1999b. Do North Atlantic bluefin tuna spawn in the Mid-Atlantic? Results of pop-up satellite tagging on spawning size class fish in the Gulf of Maine. ICCAT, Coll. Vol. Sci. Pap. SCRS/98/76

MATHER, F.J. III., and M.R. Bartlett. 1962. Bluefin tuna concentration found during a longline exploration of the northwestern Atlantic slope. Comm. Fish.Rev. 24:1-7.

MATHER, F.J., III, J.M. Mason, Jr., and A. C. Jones. 1995. Historical Document: Life history and fisheries of Atlantic bluefin tuna. NOAA Technical Memorandum NMFS-SEFSC-370.

RESTREPO, V. 1996. A research planning workshop for Atlantic bluefin tuna tagging studies. ICCAT, Coll. Vol. Sci. Pap. 5(2):267-278

RICHARDS, W.J. 1976. Spawning of bluefin tuna (*Thunnus thynnus*) in the Atlantic Ocean and adjacent seas. ICCAT, Coll. Vol. Sci. Pap. 5(2):267-278.

6

Exhibit I
SCRS Report on Tuna Tagging

RIVAS, L.R. 1954. A preliminary report on western north Atlantic fishes of the family Scombridae. Bull. Mar. Sci. 1:209-230.

TIEWS, K. 1963. Species Synopsis No. 13, FAO Fisheries Biology Synopsis No. 56, Synopsis of biological data on bluefin tuna Thunnus thynnus (Linnaeus) 1758 (Atlantique et Mediterranee). Rome.

WELCH, D.W. and J. P. Eveson. 1999. An assessment of light-based geoposition estimates from archival tags. Can J. Aquat. Sci 56:1317-1327.

Table 1. Atlantic bluefin tuna, *Thunnus thynnus*, tag reporting locations of fish released with pop-up satellite tags in the NW Atlantic, 11 August - 8 Oct, 1999.

ID	Release date	Gear type	Tag type	Release latitude °N	Release longitude °W	Reporting date	Reporting latitude °N	Reporting longitude °W	Length (cm) SFL	Year class Y^m
999	08/11/99	rod&reel	a	42.13	70.32	08/16/99	42.67	69.47	234	10+
29	09/20/99	rod&reel	a	41.23	69.24	10/04/99	42.67	69.47	234	10+
126	09/29/99	handline	a	42.15	65.60	10/15/99	41.82	64.05	203	8^9
147	09/25/99	rod&reel	a	41.23	69.24	10/24/99	41.08	66.59	178	7^1
212	10/08/99	p. seine	a	41.29	69.09	12/25/99	34.65	70.27	200	8^7
216	10/08/99	p. seine	a	41.29	69.09	12/25/99	39.54	56.33	211	9^4
106	07/12/99	rod&reel	s	42.45	70.30	03/09/20	36.70	42.14	241	10+
215	10/08/99	p. seine	a	41.29	69.09	04/14/20	33.86	40.29	211	9^4
211	10/08/99	p.seine	a	41.29	69.09	04/14/20	29.55	52.89	224	10+
202	09/27/99	rod&reel	a	41.30	69.24	04/14/20	37.93	37.85	224	10+
205	09/25/99	rod&reel	a	41.29	69.22	04/14/20	54.63	21.36	266	10+
101	07/29/99	rod&reel	s	42.28	69.38	05/11/20	40.25	43.51	191	7^11
204	09/24/99	rod&reel	a	41.28	69.21	06/30/20	33.65	59.67	193	8^1
206	09/24/99	rod&reel	a	41.26	69.26	06/30/20	35.85	41.01	233	10+
217	10/08/99	p. seine	a	41.29	69.09	09/01/20	36.47	38.86	221	10+
214	10/08/99	p. seine	a	41.29	69.09	09/01/20	33.32	50.39	203	8^9

Abbreviations a=popup archival tag, s=single point popup tag, SFL=straight fork length. Year class (year and month) . Assignment based on NMFS size at age table where 10+ is maximum age. All size and ages are estimated.

Table 2. Ambient temperatures recorded hourly by popup archival tags on spawning size class bluefin tuna, *Thunnus thynnus*, with tags reporting in December, 1999, and April, 2000.

Fish ID	Number of observations	Mean (°C) ± Std. Dev.	Minimum (°C)	Maximum (°C)	Mode (°C)	Release Date
205	1,872	15.3 ± 4.98	9.0	27.2	10.8	04/15/00
202	3,593	18.8 ± 2.01	11.6	24.8	18.6	04/15/00
211	3,968	22.4 ± 1.38	12.8	23.2	19.2	04/15/00
215	3,384	20.8 ± 1.91	13.8	27.3	19.7	04/15/00
206	5,532	16.1 ± 4.47	5.8	28.2	17.1	12/25/99
204	4,350	20.3 ± 2.82	11.6	28.4	18.9	12/25/99

Exhibit I

SCRS Report on Tuna Tagging

STEPHEN SLOAN
SUITE 1512
230 PARK AVENUE
NEW YORK. NY 10169

TEL 212 • 688 • 7567
FAX 212 • 751 • 1384
INTERNET: fishsave●pipeline.com

June 28,1999

Mr. John Sawhill
Chairman
the H. John Heinz III Center for Science, Economics and the Environment
1001 Pennsylvania Avenue, NW.
Suite 735 South
Washington, DC. 20004

Dear Mr. Sawhill,

I just received your pamphlet entitled "Managing U.S. Marine Fisheries".
I noticed that no attention has been paid to recreational fishermen in your planning
to assist the non-partisan policy options for consideration during the Magnuson-Stevens
Fishery Conservation and Management Act. Of the Collaborator Team and Senior
Advisors no one represents Recreational Fishermen. I have also checked your web site and
found that you do economically value recreational fishermen with "64 million fishing trips
generating an additional $15 billion."

I am enclosing my resume via my fishing card. I ask that I be placed on your
Collaborating Team representing recreational fishermen. I am uniquely qualified to assist
you in this important work.
I am enclosing a commendation I received as Chairman
of MAFAC which was in total disarray before my stewardship.

In addition I have a weekly nationwide radio program dedicated to fishery issues.
Many scientists,authors and environmentalists have been on the programs.
The List includes:
Dr. Sylvia Earle [the first radio show done under the sea on a coral reef]
Dr. Rebecca Lent from HMS at NMFS
Dr. Ellen Prager Scientist and in charge of YOTO and the Drifter Educational Program
Dr. Andy Kemmerer Head of the Gulf office of NMFS
Dr. Felicia Coleman reported on her research on Gag Groupers and Reefs [suggested by
                   the Environmental Defense Fund]
Dr. Jo Anne Buckholder[The Waters Turned to Blood] report on Phisteria
Dr. Anne Platt McGinn [Report on our Oceans]

Rollie Schmitten has been on three times on various issues for NMFS
Michael Leech President of IGFA [International Game Fish Associated]

Exhibit J

Sloan Letter to Heinz Center of June 28, 1999

page 2-

Sebastian Junger[The Perfect Storm], William Broad [The Universe Below and two time Pulitzer Prize winner at the New York Times], Dr. Sylvia Earle second time[Sea Change], George Reiger [The Striped Bass Chronicles],
Russell Drum [In the Slick of the Cricket] Shark Fishing Eastern Long Island,
The Japanese Whaling Commission, General Mills regarding PETA., The Whirling Disease Foundation, Wild Salmon Center, ICCAT officials, Rich Ruias[The East Coast Tuna Association], Project Access [for handicapped fishermen] and many others from all areas of fishery issues including a show several years ago by the Virginia Chapter of the Nature Conservatory.

This coming Saturday will be my 382nd consecutive live broadcast. I would like to invite you on the show. You can explain the mission of the Heinz Center. The show can be done from your home by calling our troll free number 1-800-298-8255.
We broadcast to 80 stations and reach over 600 cities every Saturday morning at 6:10 a.m.

I would be delighted to come to Washington to discuss the contents of this letter.

Tight Lines and a Good Drift

Sincerely,

Stephen Sloan

Exhibit J

Sloan Letter to Heinz Center of June 28, 1999

The H. John Heinz III Center
for Science, Economics
and the Environment

12 July 1999

Stephen Sloan
Suite 1512
230 Park Place
New York, New York 10169

Dear Mr. Sloan:

Thank you for your letter expressing interest in The Heinz Center's Managing U.S. Marine Fisheries program. I am pleased to be able to reassure you that we have indeed paid attention to recreational fisheries in our program and that, as the acknowledgement of recreational fishing on our website indicates, we are well aware of their value.

We have finished the first phase of our program, which involved interviewing a large number of people involved in U.S. fisheries and fishery management. The product of the first phase is a book to be published in early 2000 by Island Press, and I think you will be pleased to see recreational fisheries and their management included throughout the text. In addition, a paper summarizing the perspectives of those we interviewed will be posted on our website later this year. In the second phase we are conducting a series of meetings with people active in the eight regional fishery management councils. Recreational fisheries will be an integral part of these discussions.

With regard to your kind invitation to appear on your radio program, we would be happy to consider the possibility for early 2000, when the reports of our program's work will be released.

Sincerely yours,

Susan Hanna

Susan Hanna
Program Manager
Managing U.S. Marine Fisheries

cc: Merrell
        Katsouros
        Friedman
        Hanna

1001 PENNSYLVANIA AVENUE, N.W., SUITE 735 SOUTH, WASHINGTON, D.C. 20004
TELEPHONE (202) 737–6307   TELEFAX (202) 737–6410

Exhibit J
Reply Letter of Heinz to Sloan of July 12, 1999

Economic data
on sport fishing throughout the
entire United States.

Exhibit K
The Economic Importance of Sport Fishing

RECREATIONAL FISHING IS MUCH MORE THAN A TRADITIONAL AMERICAN PASTIME.

That same activity is also an immensely powerful part of our collective economic fabric, creating nearly 1.2 millions jobs nationwide. New studies now show that annual spending by America's 35.2 million adult anglers (16 years old and older) amounts to a whopping $37.8 billion. By comparison, and if hypothetically ranked as a "corporation," that revenue figure would put sport fishing in thirteenth place on the Fortune 500 list of America's largest businesses, ranking just above such global giants as Texaco and DuPont.

The impact on the American economy of all that spending is extraordinary. When that spending figure was "crunched" recently by economists at the American Sportfishing Association (ASA) to account for "ripple" or economic-multiplier effects, anglers' annual spending was shown to have:

Cover illustration Ron Barrett

- Created a nationwide economic impact of about $108.4 billion.

- Supported 1.2 million jobs, or slightly more than 1 percent of America's entire civilian labor force, in all sectors of the American economy.

- Created household income (salaries and wages) totaling $28.3 billion, which is roughly equivalent to almost half of America's entire military payroll.

- Added $2.4 billion to state tax revenues, or nearly 1 percent of all annual state tax revenues combined.

- Contributed $3.1 billion in federal income taxes, which equates to nearly a third of the entire federal budget for agriculture.

Exhibit K
The Economic Importance of Sport Fishing

ANGLER SPENDING INCREASED IN 1996

Those remarkable 1996 spending numbers are derived from the latest United States Fish and Wildlife Service (USFWS) National Survey of Fishing, Hunting and Wildlife Associated Recreation, conducted every five years in conjunction with the U.S. Census Bureau. Economic analysts for the American Sportfishing Association (ASA; through its economics program) have then used sophisticated computer models to determine the effects of that spending nationwide.

Aggregate spending on sport fishing increased in real dollars by an amazing 36 percent in the five years since the last survey was taken in 1991, when the total was $27 billion (adjusted for inflation to 1996 dollars). The American economy has been on a roll since emerging from a mild recession in 1991. Jobs, wages, and taxes generated by that spending have all shown corollary increases (see chart comparing 1991 to 1996).

Sport fishing isn't alone in showing those kinds of

Aggregate spending on sport fishing increased in real dollars by an amazing 36% in the five years since the last survey.

increases as a buoyant economy, for example, has generated widespread demand for durable goods. Annual sales of cars, trucks, and buses nationwide increased by almost 40 percent from 1991 to 1994, as just one example. And the overall installment-credit load carried by American consumers—meaning loans used to finance things such as washing machines and garden tractors, as well as camping trailers and fishing boats—increased by about 39 percent from 1990 to 1995.

When times are good, people spend money. And 35.2 million of those people, or roughly one in six Americans age 16 and older, choose to spend a great

Comparison *of* Sport Fishing Economic Impact *in* 1991 *and* 1996

	1991	1996
Expenditures	$27.6 billion	$37.7 billion
Overall Economic Impact	$79.8 billion	$108.4 billion
Wages and Salaries	$22.0 billion	$28.2 billion
Jobs	924,600	1,210,100
State Sales Tax	$1.2 billion	$1.9 billion
State Income Tax	$261.1 million	$450.6 million
Federal Income Tax	$2.4 billion	$3.0 billion

Economic figures for 1991 were adjusted for inflation to 1996 dollars.

2

Exhibit K
The Economic Importance of Sport Fishing

deal of that money on sport fishing. Part of the increase, too, comes from more time spent on the water. America's anglers spent 22 percent more time fishing in 1996 than in 1991, according to the USFWS survey.

As individuals, anglers may not spend a great deal of money. The overall average is about $1,100 per person per year, which can include everything from spending a few bucks at the local bait shop to the cost of a distant bass-fishing trip. Some spend considerably less; others much more. But at 35 million strong, anglers add up to a major economic force (nearly 50 million when anglers under 16 are included). Neither golf nor tennis—with 23.1 and 11.5 million U.S. players respectively— even comes close to that number.

Of the $37.8 billion total spent by anglers in 1996, $15.4 billion was for travel-related costs while another $19 billion went for equipment ranging from reels and fishing lures to sport-utility vehicles and boats. Another $2.3 billion was spent on land leases or land ownership for fishing, and about $570 million went for fishing licenses, permits, and fees.

RODS, REELS, LURES, CLOTHES, BOATS AND MORE – AN ANGLER'S EXPENDITURES QUICKLY ADD UP.

THE MAGIC OF MULTIPLIERS

While the spending figures are impressive by themselves, they become even more so through consideration of "ripple" or multiplier effects. This is basically the modern economist's way of saying that

Money is like manure because the more it's spread around the more good it does.

money is like manure because the more it's spread around the more good it does. Each dollar spent by an angler increases another person's income, enabling that person (or business) to spend more, which in turn increases income for somebody else. The process continues as a wide series of ripples through local, regional, and national economies until the spreading fragments of the original dollar become so small they can no longer be measured.

If enough money's spent—and remember that anglers collectively spend billions—businesses that benefit from the rippling cycle might have to add employees whose wages, when spent, will support still more jobs. Taxes on both sales and income will also be generated at greater amounts in most cases.

Within the fishing-tackle industry itself, ASA represents more than 600 organizations in the sport fishing industry, including many fish-and-game agencies, media groups, and angler organizations. And that same industry nationwide supports about 2,400

3

Exhibit K
The Economic Importance of Sport Fishing

wholesalers and distributors, 6,000 fishing tackle shops, 3,800 sporting-goods stores, about 1,000 marine dealers, and more than 6,000 other retailers ranging from department stores to catalog showrooms. But it's

Number of Anglers and Fishing Participation in 1996

State	Total anglers	Fresh water anglers	Salt water anglers	Great Lakes anglers	Total fishing days	Fresh water fishing days	Salt water fishing days	Great Lakes fishing days
Alabama	984,202	842,740	159,576		16,532,936	14,255,595	1,560,828	
Alaska	462,988	313,137	282,620		5,331,346	3,601,654	1,948,853	
Arizona	512,167	483,337			4,689,326	4,689,326		
Arkansas	763,878	739,489			9,661,166	9,661,166		
California	2,721,829	2,174,536	1,049,208		36,913,648	28,986,921	7,301,523	
Colorado	829,800	787,053			8,232,349	8,232,349		
Connecticut	418,558	317,746	177,959		5,482,553	3,880,197	1,746,733	
Delaware	196,027	65,763	147,687		2,509,479	979,923	1,611,970	
Florida	2,864,021	1,137,240	2,255,171		45,464,797	18,409,121	25,139,988	
Georgia	1,087,477	967,477	137,463		15,171,009	12,857,261	992,956	
Hawaii	260,005	22,378	243,647		3,054,915	188,688	2,901,414	
Idaho	483,459	473,587			4,411,073	4,411,073		
Illinois	1,351,047	1,123,401		259,642	20,459,075	17,089,197		1,542,490
Indiana	992,420	862,646		59,616	15,810,948	13,465,419		786,862
Iowa	496,841	476,909			7,062,085	7,062,085		
Kansas	363,992	341,001			6,355,165	6,355,165		
Kentucky	816,584	772,046			9,630,958	9,630,958		
Louisiana	1,031,141	814,538	346,016		20,987,388	18,492,978	2,082,839	
Maine	356,382	289,832	106,309		5,114,006	4,107,475	989,127	
Maryland	714,911	318,606	497,826		10,194,931	4,289,686	5,264,323	
Massachusetts	703,713	377,423	429,101		10,133,716	6,746,413	3,953,464	
Michigan	1,823,534	1,310,695		674,200	28,708,731	19,456,110		6,084,370
Minnesota	1,538,180	1,420,525		48,274	27,002,340	25,896,941		162,718
Mississippi	578,985	487,463	121,102		9,731,610	8,212,938	1,442,835	
Missouri	1,208,793	1,137,764			14,681,924	14,681,924		
Montana	335,484	328,827			2,617,100	2,617,100		
Nebraska	269,119	246,980			3,004,392	3,004,392		
Nevada	223,745	218,505			1,975,947	1,975,947		
New Hampshire	267,338	237,096	45,740		3,540,689	3,138,868	314,057	
New Jersey	1,058,672	427,913	841,372		16,125,449	6,020,671	10,366,335	
New Mexico	321,373	311,740			2,836,025	2,836,025		
New York	1,706,048	1,110,693	476,203	415,413	29,359,346	17,411,647	5,151,077	6,419,287
North Carolina	1,556,655	1,008,740	769,614		22,229,920	15,831,196	5,676,942	
North Dakota	97,496	89,867			1,320,936	1,320,936		
Ohio	1,231,445	907,729		452,555	17,848,215	12,877,546		3,539,390
Oklahoma	924,149	890,889			14,673,615	14,673,615		
Oregon	657,919	588,928	161,899		7,989,172	7,117,488	870,178	
Pennsylvania	1,354,873	1,277,461		83,821	20,901,248	18,635,326		709,366
Rhode Island	163,095	71,677	107,834		2,154,866	1,347,100	947,116	
South Carolina	986,392	715,778	381,714		15,018,068	11,341,482	2,451,423	
South Dakota	227,394	213,404			2,747,868	2,747,868		
Tennessee	859,679	766,582			11,316,579	11,316,579		
Texas	2,612,743	2,146,758	861,738		51,328,728	37,575,122	13,030,173	
Utah	405,943	397,023			3,926,202	3,935,544		
Vermont	188,290	175,971			1,951,173	1,951,173		
Virginia	1,029,100	761,067	376,889		14,570,507	9,281,915	5,155,718	
Washington	1,005,044	768,209	378,404		12,859,716	10,975,174	2,134,453	
West Virginia	335,584	322,824			5,039,979	5,039,979		
Wisconsin	1,473,561	1,232,103		180,652	17,130,434	14,397,635		850,474
Wyoming	413,141	378,549			2,414,943	2,414,943		
US TOTAL	35,245,809	28,920,776	9,438,238	2,038,797	625,892,832	485,473,711	103,034,328	20,094,956

Exhibit K
The Economic Importance of Sport Fishing

important to remember that the 1.2 million jobs supported in 1996 by anglers having spent $37.8 billion aren't just jobs in sporting-goods stores, but can

Anglers flock to area rivers—and spend money—in early spring, long before other forms of summer tourism bring added dollars to the region.

include everyone from health-clinic employees to your local telephone repairman. Here's a specific example of just how that works.

If you fish in southern Wisconsin, for example, you might have stopped in at the Ace Hardware store in La Crosse. Here sales clerk Ron Gerhke is liable to suggest some RC Buzzbait lures for the local, largemouth-bass fishing. So you plunk down $10 for a trio of likely lures and head happily for the nearest bass pond. Then that $10 starts a ripple effect, spreading outward just like the ripples made when your lure hits the water.

ANGLERS SPEND $6.4 BILLION ANNUALLY ON BOATS, PARTS AND ACCESSORIES SPECIFICALLY TO GO FISHING.

Part of that money goes into Ron's wages, helping to buy clothes for his kids at the local Farm & Fleet store. Part goes for income taxes, and yet another part goes into the store's overhead, paying for things like the electric bill from Northern States Power. And part of that money goes to Bettendorf, Iowa, where Ryan Coon of RC Tackle has a part-time business

assembling lures in the family basement. Ryan pays bills, too, of course, and the rippling cycle further spreads and repeats. Included therein is money for basic family needs such as health care and telephone repair, which is how the effect of your tackle purchase spreads far beyond the doors of a sporting-goods store.

Ten dollars isn't very significant, of course, but when 35 million anglers spend $37.8 billion in 12 months the result in jobs, wages, and other economic effects is both extraordinary and at the very foundation of America's economic health.

Suppose that sport fishing were eliminated overnight. That doesn't mean the 1.2 million jobs it supports would evaporate. Former anglers would spend their money on other things such as bowling or video games, and economic multiplier effects would still take place, although in different directions. Stores and businesses that catered to anglers would go belly up, and their employees would find different jobs. Or so the theory goes.

Sport fishing means, jobs, wages, and dinner on the table for tens of thousands of families in areas where alternatives are often limited.

This switch is called "convertibility," and it's an argument often used against anyone who cites the

5

Exhibit K
The Economic Importance of Sport Fishing

importance of specific economic impacts. But for sport fishing, that argument doesn't wash. Here's why.

added dollars to the region, thereby filling another seasonal gap.

Convertibility, meaning job changes or alternative businesses, takes place most readily in cities where the range of alternatives is greatest. Recreational fishing, on the other hand, takes place most often in rural areas where job alternatives are typically few or none. Here anglers' dollars often mean the difference between real jobs or hard-core unemployment. That rural distinction is further emphasized by sport fishing's often-seasonal nature. In New York's Catskill Mountains and in northern Michigan, as a couple of examples, trout anglers flock to area rivers—and spend money—in early spring, long before other forms of summer tourism bring

Angler Trip *and* Equipment Expenditures *in the* United States *for* 1996

Expenditure Item	Total
TRIP EXPENDITURES	
Food, Drink and Refreshments	$ 4,255,842,791
Lodging	$ 1,733,823,092
Public Transportation	$ 559,029,278
Private Transportation	$ 3,171,216,027
Boat Fuel	$ 1,339,584,467
Guide Fees, Pack Trip or Package Fees	$ 638,466,383
Public Land Use or Access Fees	$ 140,258,431
Private Land Use or Access Fees	$ 84,353,614
Boat Launching Fees	$ 201,377,081
Boat Mooring, Storage, Maintenance and Insurance	$ 1,398,154,895
Equipment Rental	$ 331,308,320
Bait (live, cut, prepared)	$ 1,084,661,194
Ice	$ 319,258,420
Heating and Cooking Fuel	$ 123,883,241
FISHING EQUIPMENT EXPENDITURES	
Rods, Reels, Poles and Rod Making Components	$ 2,331,835,635
Lines and Leaders	$ 490,917,008
Artificial Lures, Flies, Baits and Dressing	$ 880,910,433
Hooks, Sinkers, Swivels, etc.	$ 376,671,950
Tackle Boxes	$ 128,193,348
Creels, Stringers, Fish Bags, Landing Nets and Gaff Hooks	$ 95,915,440
Minnow Traps, Seines and Bait Containers	$ 66,220,786
Depth Finders, Fish Finders and Other Electronic Fishing Devices	$ 395,926,970
Ice Fishing Equipment	$ 97,557,372
Other Fishing Equipment	$ 444,526,129
AUXILIARY PURCHASES FOR FISHING	
Camping Equipment	$ 501,711,047
Binoculars, Field Glasses, Telescopes, etc.	$ 46,757,879
Special Fishing Clothing, Foul Weather Gear, Boots, Waders, etc.	$ 312,636,188
SPECIAL EQUIPMENT PURCHASED FOR FISHING	
Bass Boat	$ 2,005,235,791
Other Motor Boat	$ 3,220,523,391
Canoe or Other Non-Motor Boat	$ 144,712,414
Boat Motor, Boat Trailer/Hitch or Other Boat Accessories	$ 981,703,104
Pickup, Camper, Van, Travel or Tent Trailer, Motor Home, House Trailer	$ 4,573,214,215
Cabin	$ 27,394,985
Trail Bike, Dune Buggy, 4x4 Vehicle, 4-Wheeler, Snowmobile	$ 1,129,232,231
Other Special Equipment Including Ice Chest	$ 746,301,786
OTHER EXPENDITURES	
Fishing License Fees	$ 519,060,780
Other Fees	$ 60,691,571
Owned or Leased Property	$ 2,340,344,488
Processing and Taxidermy Costs	$ 62,019,727
Books and Magazines	$ 169,546,449
Dues or Contributions to Organizations	$ 152,447,837
Other Purchases	$ 113,635,846
UNITED STATES TOTAL	**$ 37,797,062,032**

Exhibit K
The Economic Importance of Sport Fishing

TOURISM-BASED ECONOMIES DEPEND ON SPORT FISHING

There are abundant examples both large and small, partly because economic impacts can be studied at

Industries such as power, timber, and agriculture have traditionally pitted the need to protect their jobs against healthy fisheries.

local and state levels as well as nationally. Lake Fork in Texas, for example, is a reservoir with superlative bass fishing that draws anglers from all over the country. According to a recent study by Texas A&M University, anglers there spend a whopping $27 million a year in the immediate, tri-county area on things such as motels, groceries, gasoline, and marina fees. More than half of that money comes in from

TRAVEL, LODGING, FOOD AND MORE – THE TRAVEL AND TOURISM SECTOR DIRECTLY GAINS $15 MILLION ANNUALLY FROM SPORT FISHING THAT TRANSLATES INTO 523,000 JOBS.

outside the immediate area, creating jobs and benefiting families there in ways that might not otherwise be possible. From King Salmon, Alaska, to

Greenville, Maine, to Flamingo, Florida, the story's much the same. Sport fishing means, jobs, wages, and dinner on the table for tens of thousands of families in areas where alternatives are often limited. Clearly, better fishing is better business, too.

That also helps to point out the importance of sport fishing tourism as an economic force. A recent study in Massachusetts, as another example, showed that about $93 million a year was being spent in that state by nonresident marine anglers–an amount that was supporting 3,300 local jobs. If sport-fishing tourism in the United States was likewise promoted to the 25 million anglers in Western Europe or the 20 million anglers in Japan, the boost to state and national economies here could be extraordinary.

EFFECTS AT DIFFERENT LEVELS

Economic effects can be especially important when calculated at the state level, and we've included a chart showing the impacts of 1996 angler spending in your state. In California, for example, that added up to $3.3 billion, supporting 74,420 jobs, creating $1.9 billion in

Top Ten States Ranked *by* Overall Economic Impact						
State	Angler Expenditures	Overall Economic Impact	Salaries and Wages	Jobs	State Sales and Income Taxes	Federal Income Tax
California	$3,324,359,199	$7,127,585,206	$1,912,882,755	74,420	226,612,888	$214,031,472
Texas	$2,869,558,423	$6,366,580,439	$1,647,197,166	80,282	179,347,401	$168,271,354
Florida	$3,288,844,296	$6,057,317,747	$1,711,404,281	81,815	197,330,658	$176,392,657
Minnesota	$1,874,835,053	$3,678,165,611	$ 948,349,442	47,293	160,177,150	$ 95,957,163
Illinois	$1,568,471,459	$3,618,451,181	$ 975,473,066	40,005	126,092,920	$106,974,691
New York	$1,785,947,624	$3,123,990,172	$ 720,283,674	28,351	92,540,895	$ 80,260,311
North Carolina	$1,571,726,554	$2,997,403,521	$ 776,525,926	40,319	93,613,732	$ 76,968,519
Michigan	$1,506,227,841	$2,854,443,939	$ 772,711,715	35,579	120,615,210	$ 80,976,932
Georgia	$1,121,277,910	$2,290,557,133	$ 615,582,528	27,808	72,052,251	$ 64,543,462
Wisconsin	$1,072,569,520	$2,137,500,309	$ 565,969,487	30,410	75,223,729	$ 61,001,694

Exhibit K

The Economic Importance of Sport Fishing

household income (salaries and wages), and creating $226.7 million in state tax revenues. And in Florida, meanwhile, some 2.9 million anglers spent $3.3 billion and supported nearly 82,000 jobs. Those are the kinds of numbers that should turn the head of any state legislator or politician. And that's just the point.

Economic Impact of Sport Fishing in the United States by State for 1996

State	Angler Expenditures	Overall Economic Impact	Salaries and Wages	Jobs	State Sales Tax	State Income Tax	Federal Income Tax
Alabama	$ 835,615,325	$1,640,836,023	$ 440,249,304	22,917	$33,424,613	$16,168,180	$ 43,565,854
Alaska	$ 548,364,219	$ 956,793,847	$ 261,571,620	12,626	no tax	no tax	$ 26,843,763
Arizona	$ 358,143,614	$ 662,936,279	$ 185,661,304	9,325	$ 17,907,181	$ 3,975,239	$ 18,705,908
Arkansas	$ 301,828,952	$ 584,559,776	$ 154,045,789	9,080	$ 13,582,303	$ 4,744,633	$ 14,201,710
California	$3,324,359,199	$7,127,585,206	$1,912,882,755	74,420	$199,461,552	$27,151,336	$214,031,472
Colorado	$ 634,446,791	$1,315,893,039	$ 358,525,912	17,835	$ 19,033,404	$ 1,815,652	$ 36,313,033
Connecticut	$ 284,277,505	$ 522,234,595	$ 145,548,006	5,562	$ 17,056,650	$ 6,174,109	$ 16,374,409
Delaware	$ 276,732,808	$ 438,751,045	$ 104,030,839	5,220	no tax	$ 4,117,074	$ 10,470,569
Florida	$3,288,844,296	$6,057,317,747	$1,711,404,281	81,815	$197,330,658	no tax	$176,392,657
Georgia	$1,121,277,910	$2,290,557,133	$ 615,582,528	27,808	$ 44,851,116	$27,201,134	$ 64,543,462
Hawaii	$ 130,038,758	$ 238,088,176	$ 70,046,235	3,080	$ 5,201,550	$ 3,961,275	$ 7,472,825
Idaho	$ 279,949,546	$ 461,681,805	$ 116,552,240	6,884	$ 13,997,477	$4,627,777	$ 10,711,682
Illinois	$1,568,471,459	$3,618,451,181	$ 975,473,066	40,005	$ 98,029,466	$28,063,454	$106,974,691
Indiana	$ 799,252,121	$1,677,490,348	$ 437,402,937	21,042	$ 39,962,606	$14,156,081	$ 44,903,333
Iowa	$ 338,969,069	$ 654,502,272	$ 171,570,996	9,118	$ 16,948,453	$ 7,936,862	$ 16,785,965
Kansas	$ 180,018,571	$ 356,981,567	$ 85,216,003	4,922	$ 8,820,910	$ 2,108,149	$ 7,954,256
Kentucky	$ 517,028,662	$1,046,748,929	$ 267,612,640	14,082	$ 31,021,720	$12,673,532	$ 26,347,451
Louisiana	$ 824,339,739	$1,546,264,215	$ 406,206,498	21,507	$ 32,973,590	$ 8,074,688	$ 39,766,936
Maine	$ 348,548,103	$ 568,029,246	$ 145,587,916	8,641	$ 20,912,886	$ 3,550,541	$ 13,316,094
Maryland	$ 475,266,219	$ 896,205,153	$ 246,766,760	11,361	$ 23,763,311	$ 9,270,742	$ 25,858,098
Massachusetts	$ 524,574,998	$ 988,599,888	$ 274,805,177	11,219	$ 26,228,750	$14,881,649	$ 30,168,684
Michigan	$1,506,227,841	$2,854,443,939	$ 772,711,715	35,579	$ 90,373,670	$30,241,540	$ 80,976,932
Minnesota	$1,874,835,053	$3,678,165,611	$ 948,349,442	47,293	$121,864,278	$38,312,872	$ 95,957,163
Mississippi	$ 703,691,648	$1,305,817,664	$ 299,881,501	16,971	$ 49,258,415	$ 5,511,595	$ 28,325,079
Missouri	$ 702,977,501	$1,445,273,434	$ 371,866,985	19,540	$ 29,700,799	$11,818,736	$ 36,558,633
Montana	$ 243,500,824	$ 447,974,606	$ 123,422,673	7,505	no tax	$ 214,788	$ 11,114,641
Nebraska	$ 235,814,547	$ 426,679,493	$ 117,629,892	6,448	$ 11,790,727	$ 3,068,318	$ 11,291,237
Nevada	$ 211,092,356	$ 335,701,417	$ 92,246,647	4,240	$ 13,721,003	no tax	$ 9,684,969
New Hampshire	$ 320,449,283	$ 580,470,601	$ 164,377,189	7,710	no tax	no tax	$ 17,084,785
New Jersey	$1,025,230,011	$2,029,864,199	$ 566,132,532	21,910	$ 61,513,801	$ 7,989,864	$ 63,342,240
New Mexico	$ 195,011,883	$ 343,812,168	$ 81,930,905	4,797	$ 9,750,594	$ 1,040,627	$ 7,574,636
New York	$1,785,947,624	$3,123,990,172	$ 720,283,674	28,351	$ 71,437,905	$21,102,990	$ 80,260,311
North Carolina	$1,571,726,554	$2,997,403,521	$ 776,525,926	40,319	$ 62,869,062	$30,744,670	$ 76,968,519
North Dakota	$ 83,415,107	$ 148,467,067	$ 37,944,621	2,252	$ 4,170,755	$ 485,951	$ 3,471,081
Ohio	$ 836,191,596	$1,879,177,292	$ 494,140,930	22,639	$ 41,809,580	$10,531,308	$ 51,865,680
Oklahoma	$ 490,767,292	$1,012,537,832	$ 258,906,659	14,797	$ 22,084,528	$ 5,472,069	$ 24,252,897
Oregon	$ 622,806,450	$1,173,234,473	$ 304,891,215	14,940	no tax	$16,316,641	$ 31,090,558
Pennsylvania	$ 649,762,961	$1,339,801,973	$ 357,441,359	16,677	$ 38,985,778	$10,008,145	$ 37,205,929
Rhode Island	$ 136,792,521	$ 231,350,837	$ 63,860,775	3,034	$ 9,575,476	$ 1,817,753	$ 6,610,011
South Carolina	$ 707,100,241	$1,332,272,691	$ 351,107,282	18,932	$ 35,355,012	$ 9,720,845	$ 34,001,282
South Dakota	$ 206,431,791	$ 351,939,997	$ 91,476,747	5,401	$ 8,257,272	no tax	$ 8,403,444
Tennessee	$ 474,724,071	$ 989,463,949	$ 265,237,749	12,812	$ 28,483,444	no tax	$ 27,238,685
Texas	$2,869,558,423	$6,366,580,439	$1,647,197,166	80,282	$179,347,401	no tax	$168,271,354
Utah	$ 231,291,509	$ 468,403,271	$ 124,003,524	6,773	$ 11,275,461	$ 5,193,480	$ 11,894,190
Vermont	$ 103,482,213	$ 178,061,022	$ 50,101,732	2,761	$ 4,139,289	$ 1,198,329	$ 4,793,318
Virginia	$ 821,317,778	$1,625,627,755	$ 441,752,462	20,880	$ 28,746,122	$15,461,130	$ 45,811,064
Washington	$ 704,396,393	$1,358,381,838	$ 373,822,490	16,713	$ 45,785,766	no tax	$ 39,676,035
West Virginia	$ 204,922,711	$ 308,804,127	$ 71,238,378	4,450	$ 12,295,363	$ 2,048,445	$ 6,323,516
Wisconsin	$1,072,569,520	$2,137,500,309	$ 565,969,487	30,410	$ 53,628,476	$21,595,253	$ 61,001,694
Wyoming	$ 174,575,258	$ 293,067,453	$ 72,705,885	4,670	$ 6,983,010	no tax	$ 6,322,698

Exhibit K
The Economic Importance of Sport Fishing

"Industries such as power, timber, and agriculture have traditionally pitted the need to protect their jobs against healthy fisheries," said ASA President Mike Hayden. "With this economic information in hand, conservationists can now argue for both fish and jobs."

Americans have more reasons than ever before to preserve and enhance fish habitat—an effort that greatly benefits the country as a whole.

MORE CLOUT FOR FISHERIES

While recreational fishing gets wide acknowledgment as a traditional activity based on a healthy environment, it tends to get little recognition as a major economic force locally, regionally, and nationally. A loss of local jobs because of a manufacturing plant closing gets banner headlines regionwide. When sport fishing opportunities are somehow diminished, jobs are also lost. But too often little noticed.

The economic importance of sport fishing "is a vital tool for state fish-and-wildlife agencies as they present their programs and budget needs to both politicians and the public," Hayden added. "As state fishing programs compete for money in state legislative arenas, this new economic information should give those programs some much-needed clout." In this context,

governments to look at fish-and-wildlife moneys as treasure troves they can divert to other budgetary needs, as if sport fishing was merely a frivolous concern. But this ASA analysis shows just how wrong that notion can be.

HEALTHIER FISHERIES MEAN BETTER BUSINESS, ADDITIONAL JOBS AND MORE OPPORTUNITIES FOR AMERICAN FAMILIES TO SPEND TIME TOGETHER.

state-by-state impact figures for angler spending are especially important. It's all too easy for state

North American Fisherman

Economic Impact *of* Saltwater Sport Fishing *in* 1996				
State	Angler Expenditures	Overall Economic Impact	Salaries and Wages	Jobs
Alabama	$ 124,877,005	$ 241,696,206	$ 69,203,793	4,084
Alaska	$ 243,854,311	$ 427,928,315	$ 117,681,065	5,713
California	$ 791,446,913	$ 1,794,559,207	$ 498,369,450	19,113
Connecticut	$ 92,883,479	$ 171,169,184	$ 47,421,860	1,810
Delaware	$ 158,965,679	$ 255,907,314	$ 61,767,519	3,125
Florida	$2,213,798,040	$ 4,106,397,320	$1,170,943,327	56,278
Georgia	$ 57,113,015	$ 116,690,535	$ 32,007,995	1,576
Hawaii	$ 124,477,713	$ 228,272,316	$ 67,220,958	2,948
Louisiana	$ 205,418,422	$ 395,016,185	$ 104,723,270	5,627
Maine	$ 141,570,673	$ 224,306,715	$ 55,422,239	3,205
Maryland	$ 308,313,922	$ 582,435,180	$ 159,110,412	7,291
Massachusetts	$ 221,680,025	$ 424,631,426	$ 119,005,086	4,957
Mississippi	$ 155,318,588	$ 293,416,053	$ 72,011,819	3,988
New Hampshire	$ 119,754,194	$ 223,385,525	$ 64,912,433	2,793
New Jersey	$ 746,904,429	$ 1,483,741,878	$ 414,464,135	16,112
New York	$ 557,871,957	$ 1,004,697,981	$ 248,873,305	9,633
North Carolina	$ 673,291,743	$ 1,285,277,129	$ 356,590,362	19,379
Oregon	$ 118,795,195	$ 225,167,540	$ 60,182,050	2,967
Rhode Island	$ 90,791,873	$ 154,159,762	$ 42,441,228	2,008
South Carolina	$ 161,383,144	$ 296,237,249	$ 81,381,678	4,747
Texas	$ 897,612,938	$ 1,989,532,703	$ 503,068,996	24,802
Virginia	$ 201,064,003	$ 398,927,629	$ 110,975,751	5,373
Washington	$ 201,162,776	$ 399,946,047	$ 107,815,766	4,605
US TOTAL	$8,674,748,085	$25,092,567,062	$6,659,787,684	287,707

Exhibit K
The Economic Importance of Sport Fishing

The enormously beneficial effects of anglers' spending depend on the continued or improved availability of fishing opportunities.

As this new analysis clearly shows, Americans have more reasons than ever before to preserve and enhance fish habitat—an effort that greatly benefits the country as a whole.

The American Sportfishing Association continues to advocate environmental measures that improve fish populations and supports programs that increase sport fishing participation. It is through these measures that the sport fishing industry can maintain its profitability in the long run—while helping to protect and enhance the angling heritage of all Americans.

For more information, call:
American Sportfishing Association
(703) 519-9691

Economic Impact *of* Freshwater Sport Fishing *in* 1996

State	Angler Expenditures	Overall Economic Impact	Salaries and Wages	Jobs
Alabama	$ 703,342,024	$ 1,384,675,129	$ 366,664,072	18,605
Alaska	$ 287,936,535	$ 503,970,257	$ 137,217,222	6,548
Arizona	$ 358,143,614	$ 662,936,279	$ 185,661,304	9,325
Arkansas	$ 301,828,952	$ 584,559,776	$ 154,045,789	9,080
California	$ 2,377,748,145	$ 5,011,445,460	$ 1,326,203,701	51,957
Colorado	$ 634,446,791	$ 1,315,893,039	$ 358,525,912	17,835
Connecticut	$ 182,091,459	$ 332,781,569	$ 92,814,571	3,556
Delaware	$ 54,997,881	$ 81,155,982	$ 16,052,212	798
Florida	$ 766,725,074	$ 1,404,509,204	$ 392,357,170	18,873
Georgia	$ 1,040,657,057	$ 2,121,266,307	$ 568,404,376	25,560
Hawaii	$ 3,866,207	$ 6,820,377	$ 1,913,660	90
Idaho	$ 279,949,546	$ 461,681,805	$ 116,552,240	6,884
Illinois	$ 1,568,471,459	$ 3,618,451,181	$ 975,473,066	40,005
Indiana	$ 799,252,121	$ 1,677,490,348	$ 437,402,937	21,042
Iowa	$ 338,969,069	$ 654,502,272	$ 171,570,996	9,118
Kansas	$ 180,018,571	$ 356,981,567	$ 85,216,003	4,922
Kentucky	$ 517,028,663	$ 1,046,748,929	$ 267,612,640	14,082
Louisiana	$ 552,534,457	$ 1,031,333,307	$ 267,648,506	14,183
Maine	$ 194,576,278	$ 319,826,084	$ 83,271,939	5,093
Maryland	$ 143,578,026	$ 269,207,260	$ 74,853,680	3,461
Massachusetts	$ 274,273,777	$ 506,352,641	$ 139,070,725	5,636
Michigan	$ 1,506,227,841	$ 2,854,443,939	$ 772,711,715	35,579
Minnesota	$ 1,874,835,053	$ 3,678,165,611	$ 948,349,442	47,293
Mississippi	$ 415,674,056	$ 739,245,72	$ 154,409,701	9,309
Missouri	$ 702,977,501	$ 1,445,273,434	$ 371,866,985	19,540
Montana	$ 243,500,824	$ 447,974,606	$ 123,422,673	7,505
Nebraska	$ 235,814,547	$ 426,679,493	$ 117,629,892	6,448
Nevada	$ 211,092,356	$ 335,701,417	$ 92,246,647	4,240
New Hampshire	$ 184,466,503	$ 326,455,861	$ 89,973,309	4,452
New Jersey	$ 180,110,305	$ 353,382,940	$ 96,314,046	3,727
New Mexico	$ 195,011,883	$ 343,812,168	$ 81,930,905	4,797
New York	$ 1,181,289,958	$ 2,033,475,366	$ 447,994,792	17,844
North Carolina	$ 835,881,247	$ 1,584,173,399	$ 382,777,221	19,213
North Dakota	$ 83,415,107	$ 148,467,067	$ 37,944,621	2,252
Ohio	$ 836,191,596	$ 1,879,177,292	$ 494,140,930	22,639
Oklahoma	$ 490,767,292	$ 1,012,537,832	$ 258,906,659	14,797
Oregon	$ 453,877,941	$ 854,067,999	$ 221,361,946	10,960
Pennsylvania	$ 649,762,961	$ 1,339,801,973	$ 357,441,359	16,677
Rhode Island	$ 41,038,332	$ 68,085,198	$ 18,676,054	893
South Carolina	$ 475,855,706	$ 891,191,731	$ 229,883,660	12,229
South Dakota	$ 206,431,791	$ 351,939,997	$ 91,476,747	5,401
Tennessee	$ 474,724,071	$ 989,463,949	$ 265,237,749	12,812
Texas	$ 1,916,488,984	$ 4,228,987,664	$ 1,100,254,963	53,401
Utah	$ 231,291,509	$ 468,403,271	$ 124,003,524	6,773
Vermont	$ 103,482,213	$ 178,061,022	$ 50,101,732	2,761
Virginia	$ 594,541,682	$ 1,175,278,360	$ 315,828,306	14,819
Washington	$ 313,875,451	$ 604,680,596	$ 168,045,885	8,021
West Virginia	$ 204,922,711	$ 308,804,127	$ 71,238,378	4,450
Wisconsin	$ 1,072,569,520	$ 2,137,500,309	$ 565,969,487	30,410
Wyoming	$ 174,575,258	$ 293,067,453	$ 72,705,885	4,670
US TOTAL	$26,898,814,893	$76,919,593,071	$19,919,483,714	854,512

10

Exhibit K
The Economic Importance of Sport Fishing

Economic Value Comparison Between Only Recreational Marlin Tournaments and Yearly Commercial Marlin Fishing

Recreational Tournament Marlin Fishing
1,000 Boats x 3 days

Fuel	$ 2,880,000
Tackle	1,000,000
Dockage	1,500,000
Insurance	5,000,000
Food	600,000
Tournament Fees	3,000,000
Tournament Cost for Captain & Crew (3 per boat)	105,000,000
Telephone/Comm.	600,000
Misc. Boat/Repairs	125,000
Total Value	**$119,705,000**

Commercial Marlin Fishing
Total Commercial Fleet of 1456 boats @ 150 days per year
with 2500 hooks per set at 5% productivity factor

	M. Tons	Pounds	Value @ $1.50 per lb.
Japanese Fleet	16,500	36,300,000	$ 54,450,000
Taiwanese Fleet	20,000	44,000,000	66,000,000
European Fleet	20,000	44,000,000	66,000,000
Korea and Other ICCAT Fleet	50,000	110,000,000	165,000,000
Nonreporting Vessels IUU (345)	20,000	44,000,000	66,000,000
Total Value			**$417,450,000**

Recreational Fishing: $119,705,000 ÷ 3 (days)   = $ 39,901,667 per day
Commercial Fishing: $417,450,000 ÷ 150 (days) = $ 2,783,000.00 per day

98% of recreational fish were hooked and released and lived.
100% of commercial catch were killed

Exhibit K
Economic Value Comparison

The Fisheries Defense Fund Inc.

(A Non-Profit Organization)

Stephen Sloan - Chairman

60 East 42nd Street, Suite 1201
New York, New York 10165

TEL 212·688·7567
FAX 212·751·1384
Email: fishsave@pipeline.com

February 18, 2000

Dr. Gary Matlock
Office of Sustainable Fisheries
NOAA-NMFS
1315 East-West Highway
Silver Springs, Maryland 20910

CERTIFIED MAIL RETURN RECEIPT REQUESTED: Z363-888-751

Re: Heinz Foundation/The John Heinz III Center for Science, Economics and the Environment:

Dear Dr. Matlock,

Under the Freedom of Information Act I wish to receive copies of any grant requests made by the Heinz Foundation/ The John Heinz III Center for Science, Economics and the Environment to NMFS or NOAA for the years 1997-1998 & 1999. In addition, I wish to receive a copy of the actual grant(s) and any inter-office memorandums plus any correspondence with The Heinz Foundation for the periods of 1997-1998-1999.

Very truly yours,

The Fisheries Defense Fund Inc.

Stephen Sloan, Chairman

Stephen Sloan, Fisherman

Exhibit L

Sloan Letter to Dr. Gary Matlock of February 18, 2000

UNITED STATES DEPARTMENT OF COMMERCE
National Oceanic and Atmospheric Administration
OFFICE OF FINANCE AND ADMINISTRATION

Mr. Stephen Sloan, Chairman
The Fisheries Defense Fund, Inc.
60 East 42nd. Street, Suite 1201   APR 1 8 2000
New York, New York 10165

Reference: Freedom of Information Act Request

Dear Mr. Sloan:

This is in response to your Freedom of Information Act
(FOIA) request #2000-00193 in which you requested copies of any
grant requests made by the Heinz Foundation/The John Heinz III
Center for Science, Economics and the Environment to National
Marine Fisheries Service or NOAA for the years 1997-1999. In
addition you requested a copy of the actual grant(s), and any
inter-office memos and correspondence with the Heinz Foundation
for the periods of 1997-1999.

I have enclosed a copy of the redacted grant applications,
grant award documents, and performance reports. However, I have
determined that the information you requested regarding inter-
office memos only is exempt under 5 U.S.C. 552(b)(5), which
protects inter-agency or intra-agency memorandums or letters
which would not be available by law to a party in litigation with
the agency.

Based on the above information, this constitutes a partial
denial of your request. You may appeal the denial of information
within 30 days of receipt of this letter. Address your written
appeal to the General Counsel, Room 5882, U.S. Department of
Commerce, 14th Street and Constitution Avenue, NW, Washington,
D.C., 20240, and prominently mark your letter and the outside
envelope, "FOIA APPEAL." Your appeal should state the reasons why
you believe the requested records should be released under the
Act and why you believe this denial decision to be in error.
Attach a copy of your original request and a copy of this letter
to your appeal.

Sincerely,

Helen M. Hurcombe, Director
Acquisition and Grants Office

Enclosures

Printed on Recycled Paper

Exhibit L
Reply Letter from NOAA of April 18, 2000

April 28,2000

Ms. Helen M. Hurcombe, Director
Acquisition and Grants Office NOAA-NMFS
1325 East-West Highway
SSMC2-OF A52 Ninth Floor
Silver Spring, Maryland 20910-3283

Re: FOIA Request # 2000-00193

Dear Ms. Hurcombe,

I would like to respond to your letter of April 18,2000. I have no objection to the
redaction of proper names providing they are not presently, were at the time of the grant
request, or at the times of the grant employees of The National Marine Fisheries Service
or NOAA. If any of the redacted names fall into these mention categories I believe
you should indicate who they are and what position they held[hold] at the time.

I wish to draw your attention to Award number NA870CO312. One employee of NMFS
was on the "Collaborator Team", namely, Dr. Gary Matlock. Under this request I wish to
receive any time records that show just how much time he spent on the panel's business. A
schedule of meetings would also be in order with the attendance log of those meetings
kept by NMFS or provided to them by the Heinz Foundation. In addition, I wish to
receive the records of Mr. Richard Schaeffer's division of recreational fishing if there was
any attendance in the meetings or impute to the reports produced by the Heinz Foundation
under the grant. If not a simple statement will do.

If I receive the above information it probably will not be necessary to appeal any decisions
made in your letter of April 18,2000. I must state for the record that any appeal process
will not start until I have an answer to this letter.

Please note the new address: 510 Park Avenue,New York, NY 10022
Sincerely Yours,

The Fisheries Defense Fund Inc.

Stephen Sloan, Chairman

Stephen Sloan, Fisherman

Exhibit L

Reply Letter of Sloan to NOAA of April 28, 2000

APPLICATION FOR FEDERAL ASSISTANCE

2. DATE SUBMITTED March 25, 1998	**App** ntifier
1. TYPE OF SUBMISSION:	

1. TYPE OF SUBMISSION:		3. DATE RECEIVED BY STATE March 27	State Application Identifier
Application ☐ Construction ☑ Non-Construction	Preapplication ☐ Construction ☐ Non-Construction	**4. DATE RECEIVED BY FEDERAL AGENCY** March 27, 1998	Federal Identifier

5. APPLICANT INFORMATION

Legal Name: ~~The H. John Heinz III Center for Science, Economics & the Environment~~

Organizational Unit:

Address (give city, county, state, and zip code):
1001 Pennsylvania Avenue, NW, Suite 735 South
Washington, DC 20004

Name and telephone number of the person to be contacted on matters involving this application (give area code)
Mary Hope Katsouros
(202) 737-6307

6. EMPLOYER IDENTIFICATION NUMBER (EIN):
1 3 - 3 7 5 5 5 3 0

7. TYPE OF APPLICANT: (enter appropriate letter in box) [N]

A. State
B. County
C. Municipal
D. Township
E. Interstate
F. Intermunicipal
G. Special District

H. Independent School Dist.
I. State Controlled Institution of Higher Learning
J. Private University
K. Indian Tribe
L. Individual
M. Profit Organization Non-profit
N. Other (Specify)

8. TYPE OF APPLICATION:
☑ New   ☐ Continuation   ☐ Revision

If Revision, enter appropriate letter(s) in box(es): ☐ ☐

A. Increase Award   B. Decrease Award   C. Increase Duration
D. Decrease Duration   Other (specify):

9. NAME OF FEDERAL AGENCY:
NOAA/CSC

10. CATALOG OF FEDERAL DOMESTIC ASSISTANCE NUMBER:
1 1 - 4 7 3   (CSC)
TITLE: Coastal Services Center ~~Ecosystem Health~~

11. DESCRIPTIVE TITLE OF APPLICANT'S PROJECT:
~~U.S. Marine Fisheries Management~~
Cooperative Agreement for Coastal Stewardship between The H. John Heinz III Center for Science, Economics + the Environment + CSC.

12. AREAS AFFECTED BY PROJECT (Cities, Counties, States, etc.):
Nation

13. PROPOSED PROJECT

Start Date 05/01/98	Ending Date 09/30/99

14. CONGRESSIONAL DISTRICTS OF: DC at Large

a. Applicant DC at Large	b. Project DC at Large

15. ESTIMATED FUNDING:

a. Federal	$ 525,000 .00
b. Applicant	$ .00
c. State	$ .00
d. Local	$ .00
e. Other	$ 300,447 .00
f. Program Income	$ .00
g. TOTAL	$ 825,447 .00

16. IS APPLICATION SUBJECT TO REVIEW BY STATE EXECUTIVE ORDER 12372 PROCESS?

a. YES. THIS PREAPPLICATION/APPLICATION WAS MADE AVAILABLE TO THE STATE EXECUTIVE ORDER 12372 PROCESS FOR REVIEW ON:

DATE :

b. NO. ☑ PROGRAM IS NOT COVERED BY E.O. 12372
☐ OR PROGRAM HAS NOT BEEN SELECTED BY STATE FOR REVIEW

17. IS THE APPLICANT DELINQUENT ON ANY FEDERAL DEBT?
☐ Yes   If "Yes," attach an explanation.   ☑ No

18. TO THE BEST OF MY KNOWLEDGE AND BELIEF, ALL DATA IN THIS APPLICATION/PREAPPLICATION ARE TRUE AND CORRECT, THE DOCUMENT HAS BEEN DULY AUTHORIZED BY THE GOVERNING BODY OF THE APPLICANT AND THE APPLICANT WILL COMPLY WITH THE ATTACHED ASSURANCES IF THE ASSISTANCE IS AWARDED.

a. Type Name of Authorized Representative Mary Hope Katsouros	b. Title Vice President for Programs	c. Telephone Number (202) 737-6307
d. Signature of Authorized Representative		e. Date Signed 3/26/98

Previous Edition Usable and for Local Reproduction

Standard Form 424 (Rev. 4-92)
Prescribed by OMB Circular A-102

A-1

Exhibit L

Grant Application Forms

236

APPLICATION FOR FEDERAL ASSISTANCE	2. DATE SUBMITTED 5/26/98	Applicant Identifier	OMB Approval No. 0348-0043

1. TYPE OF SUBMISSION:

Application
- ☐ Construction
- ☒ Non-Construction

Preapplication
- ☐ Construction
- ☐ Non-Construction

3. DATE RECEIVED BY STATE — State Application Identifier

4. DATE RECEIVED BY FEDERAL AGENCY 5/27/98 — Federal Identifier

5. APPLICANT INFORMATION

Legal Name: The H. John Heinz III Center for Science, Economics & the Environment

Organizational Unit:

Address (give city, county, state, and zip code):
1001 Pennsylvania Ave., NW, Suite 735 South
Washington, DC   20004

Name and telephone number of the person to be contacted on matters involving this application (give area code)
Mary Hope Katsouros
(202) 737-6307

6. EMPLOYER IDENTIFICATION NUMBER (EIN):
1 3 - 3 7 5 5 5 3 0

7. TYPE OF APPLICANT: (enter appropriate letter in box)   N

- A. State
- B. County
- C. Municipal
- D. Township
- E. Interstate
- F. Intermunicipal
- G. Special District
- H. Independent School Dist.
- I. State Controlled Institution of Higher Learning
- J. Private University
- K. Indian Tribe
- L. Individual
- M. Profit Organization   Non-Profit
- N. Other (Specify)

8. TYPE OF APPLICATION:

☒ New   ☐ Continuation   ☐ Revision

If Revision, enter appropriate letter(s) in box(es): ☐ ☐

- A. Increase Award
- B. Decrease Award
- C. Increase Duration
- D. Decrease Duration
- Other (specify):

9. NAME OF FEDERAL AGENCY:
NOAA/Coastal Services Center

10. CATALOG OF FEDERAL DOMESTIC ASSISTANCE NUMBER:
1 1 - 4 7 3
TITLE: Coastal Services Center

11. DESCRIPTIVE TITLE OF APPLICANT'S PROJECT:
Cooperative Agreement for Coastal Stewardship Between The H. John Heinz III Center for Science, Economics and the Environment and the NOAA Coastal Services Center

12. AREAS AFFECTED BY PROJECT (Cities, Counties, States, etc.):
Nation

13. PROPOSED PROJECT

Start Date	Ending Date
06/01/98	09/30/99

14. CONGRESSIONAL DISTRICTS OF: DC at Large

a. Applicant
DC at Large

b. Project
DC at Large

15. ESTIMATED FUNDING:

	$	.00
a. Federal	275,000	.00
b. Applicant		.00
c. State		.00
d. Local		.00
e. Other	129,276	.00
f. Program Income		.00
g. TOTAL	404,276	.00

16. IS APPLICATION SUBJECT TO REVIEW BY STATE EXECUTIVE ORDER 12372 PROCESS?

a. YES. THIS PREAPPLICATION/APPLICATION WAS MADE AVAILABLE TO THE STATE EXECUTIVE ORDER 12372 PROCESS FOR REVIEW ON:

DATE _____

b. NO. ☐ PROGRAM IS NOT COVERED BY E.O. 12372
☐ OR PROGRAM HAS NOT BEEN SELECTED BY STATE FOR REVIEW

17. IS THE APPLICANT DELINQUENT ON ANY FEDERAL DEBT?
☐ Yes   If "Yes," attach an explanation.   ☒ No

18. TO THE BEST OF MY KNOWLEDGE AND BELIEF, ALL DATA IN THIS APPLICATION/PREAPPLICATION ARE TRUE AND CORRECT. THE DOCUMENT HAS BEEN DULY AUTHORIZED BY THE GOVERNING BODY OF THE APPLICANT AND THE APPLICANT WILL COMPLY WITH THE ATTACHED ASSURANCES IF THE ASSISTANCE IS AWARDED.

a. Type Name of Authorized Representative
Mary Hope Katsouros

b. Title
Vice President for Programs

c. Telephone Number
(202) 737-6307

d. Signature of Authorized Representative

e. Date Signed
5/26/98

Previous Edition Usable
Authorized for Local Reproduction

Standard Form 424 (Rev. 4-92)
Prescribed by OMB Circular A-102

A-1

Exhibit L
Grant Application Forms

ICATION FOR RAL ASSISTANCE		2. DATE SUBMITTED		Applicant Identifier		OMB Approval No. 0348-0043
OF SUBMISSION:		3. DATE RECEIVED BY STATE		State Application Identifier		
ition tion □ Preapplication □ Construction		4. DATE RECEIVED BY FEDERAL AGENCY		Federal Identifier		
n struction	□ Non-Construction					

CANT INFORMATION

ne: The H. John Heinz III Center for Science, Economics and the Environment	Organizational Unit:
give city, county, State, and zip code): 1001 Penn. Ave., NW, Ste. 735 South Washington, DC 20004	Name and telephone number of person to be contacted on matters involving this application *(give area code)* Mary Hope Katsouros (202) 737-6307

DYER IDENTIFICATION NUMBER *(EIN):*	7. TYPE OF APPLICANT: *(enter appropriate letter in box)* [N]
3 — 3 7 5 5 5 3 0	

OF APPLICATION:

□ New   ☒ Continuation   □ Revision

n, enter appropriate letter(s) in box(es) [ ] [ ]

ase Award   B. Decrease Award   C. Increase Duration
ease Duration   Other*(specify):*

A. State	H. Independent School Dist.
B. County	I. State Controlled Institution of Higher Learning
C. Municipal	J. Private University
D. Township	K. Indian Tribe
E. Interstate	L. Individual
F. Intermunicipal	M. Profit Organization
G. Special District	N. Other (Specify) __ Non-Profit Research

9. NAME OF FEDERAL AGENCY:

NOAA Coastal Services Center

LOG OF FEDERAL DOMESTIC ASSISTANCE NUMBER: [1][1] — [4][7][3] TITLE: Coastal Services Center ʸ ʸ EFFECTED BY PROJECT *(Cities, Counties, States, etc.):*	11. DESCRIPTIVE TITLE OF APPLICANT'S PROJECT: Cooperative Agreement for Coastal Stewardship between The H. John Heinz III Center for Science, Economics & the Environment and the NOAA Coastal Services Center

OSED PROJECT	14. CONGRESSIONAL DISTRICTS OF: DC at Large	
/98	Ending Date 10/31/99	a. Applicant DC at Large	b. Project DC at Large

MATED FUNDING:		16. IS APPLICATION SUBJECT TO REVIEW BY STATE EXECUTIVE ORDER 12372 PROCESS?
il	$ 250,000 ⁰⁰	a. YES. THIS PREAPPLICATION/APPLICATION WAS MADE AVAILABLE TO THE STATE EXECUTIVE ORDER 12372 PROCESS FOR REVIEW ON:
nt	$ ⁰⁰	
	$ ⁰⁰	DATE _____
	$ ⁰⁰	b. No. ☒ PROGRAM IS NOT COVERED BY E. O. 12372
	$ 183,761 ⁰⁰	□ OR PROGRAM HAS NOT BEEN SELECTED BY STATE FOR REVIEW
n Income	$ ⁰⁰	17. IS THE APPLICANT DELINQUENT ON ANY FEDERAL DEBT?
	$ 433,761 ⁰⁰	□ Yes If "Yes," attach an explanation. ☒ No

HE BEST OF MY KNOWLEDGE AND BELIEF, ALL DATA IN THIS APPLICATION/PREAPPLICATION ARE TRUE AND CORRECT, THE
ENT HAS BEEN DULY AUTHORIZED BY THE GOVERNING BODY OF THE APPLICANT AND THE APPLICANT WILL COMPLY WITH THE
IED ASSURANCES IF THE ASSISTANCE IS AWARDED.

lame of Authorized Representative pe Katsouros	b. Title Vice President for Programs	c. Telephone Number (202) 737-6307
Authorized Representative		e. Date Signed 1/28/99

Edition Usable d for Local Reproduction	Standard Form 424 (Rev. 7-97) Prescribed by OMB Circular A-102

Exhibit L
Grant Application Forms

FORM CD-451 (REV. 10-93) AO 203-26	U.S. DEPARTMENT OF COMMERCE	☐ GRANT ☒ COOPERATIVE AGREEMENT

AMENDMENT TO
FINANCIAL ASSISTANCE AWARD

ACCOUNTING CODE
9NA0100/8L2AHD00/4119
AWARD NUMBER
NA87OC0312

RECIPIENT NAME H. JOHN HEINZ III CTR. FOR SCI.,ECON.,& ENV	AMENDMENT NUMBER 1
STREET ADDRESS 1001 PENNSYLVANIA AVENUE. NW, SUITE 735 SOUTH	EFFECTIVE DATE JANUARY 1, 1999
CITY, STATE, ZIP CODE WASHINGTON, DC 20004	EXTEND WORK COMPLETION TO OCTOBER 31, 1999

DEPARTMENT OF COMMERCE OPERATING UNIT
NATIONAL OCEANIC AND ATMOSPHERIC ADMINISTRATION

COSTS ARE REVISED AS FOLLOWS:	PREVIOUS ESTIMATED COST	ADD	DEDUCT	TOTAL ESTIMATED COST
FEDERAL SHARE OF COST	$ 275,000	$ 250,000	$ -0-	$ 525,000
RECIPIENT SHARE OF COST	$ 129,276	$ 183,761	$ -0-	$ 313,037
TOTAL ESTIMATED COST	$ 404,276	$ 433,761	$ -0-	$ 838,037

REASON(S) FOR AMENDMENT

- To provide funding for the project entitled, 'Cooperative Agreement for Coastal Stewardship between the H. John Heinz III Center for Science, Economics & the Environment and the NOAA Coastal Services Center', as requested in the Recipient's application dated 1/28/99, and revision dated 4/16/99, which are incorporated by reference.

- To revise NOAA Administrative Special Award Conditions

- To incorporate Department of Commerce Financial Assistance Standard Terms & Condtions dated 10/98.

This Amendment approved by the Grants Officer is issued in triplicate and constitutes an obligation of Federal funding. By signing the three documents, the Recipient agrees to comply with the Amendment provisions checked below and attached, as well as previous provisions incorporated into the Award. Upon acceptance by the Recipient, two signed Amendment documents shall be returned to the Grants Officer and the third document shall be retained by the Recipient. If not signed and returned by the Recipient within 15 days of receipt, the Grants Officer may declare this Amendment null and void.

☒ Special Award Conditions (ATTACHMENT B [X] ADMINISTRATIVE [ ] PROGRAMMATIC)

☒ Line Item Budget (ATTACHMENT A)

☒ Other(s): 15 CFR Part 14

SIGNATURE OF DEPARTMENT OF COMMERCE GRANTS OFFICER	TITLE NOAA GRANTS OFFICER	DATE JUN 1
NAME AND SIGNATURE OF AUTHORIZED RECIPIENT OFFICIAL Mary Pope Katsouros	TITLE Senior Vice President	DATE 6/16/99

Exhibit L
Grant Application Forms

POWER BRANDS EXPANDING WORLDWIDE

○ Nearly half of Heinz's $9.4 billion in annual sales comes from outside the United States.

○ Fiscal 2000 sales volume, +3.8%.

○ Heinz's leading power brands (sales exceeding $100 million) command number-one or number-two market shares in more than 50 countries.

○ Heinz varieties are marketed in more than 200 countries and territories.

○ The *Heinz* label is one of the most powerful, global brands in the food industry, with annual sales approaching $3 billion.

○ Heinz's five fastest-growing businesses account for 65% of global sales:

Fast-Growth Businesses	FY00 Sales	3-yr Sales CAGR*
Ketchup, condiments & sauces	$1.3 billion	+7.0%
Foodservice	1.6 billion	+7.8%
Premium frozen foods	1.0 billion	+12.0%
Tuna	1.0 billion	+5.3%
Quick-serve meals	1.2 billion	+5.3%

* Compound Annual Growth Rate

Exhibit L
Pages of Heinz Annual Report of 2000

EXHIBITS

CONSOLIDATED STATEMENTS OF INCOME

H.J. Heinz Company and Subsidiaries

Jump to...

Fiscal Year Ended	May 3, 2000	April 28, 1999	April 29, 1998
(Dollars in thousands, except per share amounts)	**(53 Weeks)**	(52 Weeks)	(52 Weeks)
Sales	$ 9,407,949	$ 9,299,610	$ 9,209,284
Cost of products sold	5,788,525	5,944,867	5,711,213
Gross profit	3,619,424	3,354,743	3,498,071
Selling, general and administrative expenses	2,350,942	2,245,431	1,977,741
Gain on sale of Weight Watchers	464,617	—	—
Operating income	1,733,099	1,109,312	1,520,330
Interest income	25,330	25,082	32,655
Interest expense	269,748	258,813	258,616
Other expenses, net	25,005	40,450	39,388
Income before income taxes	1,463,676	835,131	1,254,981
Provision for income taxes	573,123	360,790	453,415
Net income	$ 890,553	$ 474,341	$ 801,566
PER COMMON SHARE AMOUNTS:			
Net income — diluted	$ 2.47	$ 1.29	$ 2.15
Net income — basic	$ 2.51	$ 1.31	$ 2.19
Cash dividends	$ 1.44 1/2	$ 1.34 1/4	$ 1.23 1/2
Average common shares outstanding — diluted	360,095,455	367,830,419	372,952,851
Average common shares outstanding — basic	355,272,696	361,203,539	365,982,290

See Notes to Consolidated Financial Statements.

go to previous page   go to next page

Exhibit L
Pages of Heinz Annual Report of 2000

Exhibit M

Report of Marlin as Cat Food by Heinz

Memorandum of Understanding and Cooperation

Parties

Blue Water Fishermen's Association
American Sportfishing Association
Coastal Conservation Association
The Billfish Foundation

Objectives

1. To achieve meaningful reductions in the bycatch and fishing mortality of undersized swordfish and all billfish species in the Atlantic and Gulf of Mexico pelagic longline fisheries within the U.S. Exclusive Economic Zone through the establishment of high migratory species conservation zones that apply time-area closures to pelagic longline fishing based on the best available scientific information.

2. To provide compensation to U.S. pelagic longline fishermen substantially adversely affected by the establishment of HMS conservation zones and the application of time-area closures through a permit buyback program.

3. To prevent adverse impacts on the economics or conservation of potentially impacted non-HMS fisheries.

4. To conduct scientific research investigating pelagic longline fishing gear and methods that reduce billfish and other highly migratory species catch and mortality while maintaining the viability of the fishery.

5. To advance the principles of small, undersized swordfish and billfish by catch reduction at ICCAT.

6. To promote a spirit of alliance, unity and mutual respect among the various sectors of the U.S. Atlantic recreational and commercial HMS fisheries.

Understanding

Recognizing the mutual interests of the parties in advancing the conservation and management of highly migratory species in the United States and internationally through ICCAT, and to achieve the aforementioned objectives, the Parties agree to the following:

Exhibit N
Memorandum of Understanding

Process

1. To implement this Memorandum in good faith and in the spirit of full cooperation, and to take such steps as are necessary to avoid and prevent any circumstances that serve to undermine or diminish such cooperation including discouraging any attempts by non-parties to negatively impact, unilaterally disadvantage and/or eliminate those Atlantic HMS fisheries engaged in by the parties.

2. To engage and apply all available political, scientific, public relations and administrative assets to implement this Memorandum in as expeditious a manner as is possible.

3. To achieve Objectives 1, 2, 3, and 4 of this Memorandum exclusively through the enactment of such authorizing and appropriations legislation as is necessary during the full term of the 106th Congress.

4. To take such actions as are necessary and appropriate to ensure that any rulemaking or other form of administrative action does not supersede, preempt or interfere with this legislative initiative.

5. To oppose any legislation and legislative amendments that are inconsistent with the objective and specific details of this Memorandum.

6. To provide or secure in a timely manner any scientific, economic or other data that is necessary for the implementation of this Memorandum.

7. To work cooperatively with Congress, the Administration, and nongovernmental entities to implement all aspects of this Memorandum.

Specifics

1. That the legislation shall establish an Atlantic swordfish/billfish conservation zone closed annually, year-round to the use of pelagic longline gear as described by the following coordinates of latitude and longitude.

(A) (North/South Carolina land border)
(B) (North/South Carolina border-seaward extension
(C) 33° 00' N/78° 00' W
(D) 33° 00' N/77° 00' W
(E) 32° 00' N/77° 00' W
(F) 32° 00' N/78° 00' W
(G) 31° 00' N/78° 00' W
(H) 31° 00' N/79° 00' W

Exhibit N

Memorandum of Understanding

(I) 24° 00' N/79° 00' W
(J) 24° 00' N/82° 00' W
(K) 25° 00' N/82° 00' W
(L) 25° 00' N/81° 00' W
(M) 26° 00' N/81° 00' W

2. That the legislation shall establish a Gulf of Mexico swordfish conservation zone closed to the use of pelagic longline gear annually from January 1st to Memorial Day as described by the following coordinates of latitude and longitude.

(A) 30° 00' N/87° 30' W
(B) 30° 00' N/86° 00' W
(C) 29° 00' N/86° 00'' W
(D) 29° 00' N/87° 30' W

3. That the legislation shall establish a Gulf of Mexico billfish conservation zone closed to the use of pelagic longline gear annually from Memorial Day through Labor Day as described by the following coordinates of latitude and longitude.

(A) 26° 00' N/97° 10' W (approx. loc. of US/Mex land border)
(B) 26° 00' N/96° 00' W
(C) 27° 00' N/94° 00' W
(D) 27° 00' N/90° 00' W
(E) 28° 00' N/90° 00' W
(F) 29° 00' N/88° 00' W
(G) 29° 00' N/86° 00' W
(H) 29° 40' N/85° 20' W (approx. loc. of Cape San Blas, Fla.)

4. That the legislation will provide for the establishment of a fishing permit "buyback" program for pelagic longline vessels to take effect at the same time as the swordfish/billfish conservation zones. The legislation will provide a process to buy all fishing permits from up to 60 eligible vessels on a willing buyer, willing seller basis. Vessels will be prevented from reflagging or fishing in any other commercial fishery including state water fisheries. Vessel owners will be compensated by paying them for all of their fishing licenses (federal and state). Vessels not documented will be named and barred from any commercial fishery. Vessels will be allowed to participate in any sector of the recreational fishery including as a charter boat.

Exhibit N

Memorandum of Understanding

5. That the legislation will provide no less than 50% of the total costs of a buyback program be paid by the Federal government through appropriations, and will provide the necessary authorities and mechanisms for half of the remaining total costs to be paid by a fee on the sale within the US of Atlantic swordfish and the other half of the remaining total costs to be paid by recreational fishermen benefiting from the closed areas.

6. That the legislation will establish a Pelagic Longline Billfish Bycatch and Mortality Reduction Research Program that identifies and tests a variety of longline gear configurations and uses in order to determine which are the most effective for reducing billfish bycatch mortality in the Gulf of Mexico yellowfin tuna fishery. The legislation shall require the precise design of this Program to be developed through a scientific workshop convened by NMFS, Southeast Fishery Science Center and that members of the pelagic longline and recreational billfish industries and their scientists will be included as part of the design team.

7. That the legislation will provide for: (i) the results of the Pelagic Longline Billfish Bycatch and Mortality Reduction Research Program to be submitted as a report to Congress no later than 3 years from the date of enactment of the legislation; (ii)the restrictions on pelagic longlining in the Gulf of Mexico billfish conservation zone to sunset no later than 4 years from the date of enactment of the legislation; and (iii) for there to be a clear nexus between the report in Congress and the BUDGET so as to ensure a comprehensive reevaluation of the most effective and practicable means to reduce billfish bycatch and billfish bycatch mortality in the yellowfin tuna pelagic longline fishery in the Gulf of Mexico.

8. That the legislation will provide the necessary authorization and appropriations to cover 100% of the costs for the Pelagic Longline Billfish Bycatch and Mortality Reduction Research Program and, in addition, at least $400,000 to the NMFS Southeast Fisher Science Center for additional billfish and associated highly migratory species research.

9. That the legislation will ensure that any future considerations of time-area closures for pelagic longline within the US EEZ are consistent with ICCAT recommendations, do not disadvantage US fishermen relative to the fishermen of other nations, and are justified by the best available scientific information. This paragraph is not intended as a vehicle to modify existing law.

Exhibit N
Memorandum of Understanding

10. That the legislation will ensure that NMFS is not precluded from conducting pelagic longline fishery research, including research involving the use of experimental pelagic longline fishing gear in any of the conservation zones at any time.

Signed:

Blue Water Fishermen's Association
Date:

American Sportfishing Association
Date: 8/16/99

Coastal Conservation Association
Date: 8/20/99

The Billfish Foundation
Date: 24. VIII. 9

Exhibit N

Memorandum of Understanding

SEA TURTLE DEMOGRAPHICS

OUTLINE

- **Introduction**
- **Incidental Takes**
- **Hook Locations**
- **Natal Origin of Loggerheads**
- **Post-Hooking Survival Pilot Study**

Exhibit O
Sea Turtle Demographics

INCIDENTAL TAKES

Species	Total	Dead/Unknown
Loggerhead	142	0
Leatherback	77	1

I02003 #007
Armpit

Exhibit O
Sea Turtle Demographics

01006 #001   Flipper and Entanglement

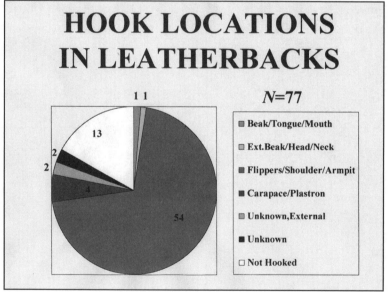

HOOK LOCATIONS IN LEATHERBACKS

N=77

- Beak/Tongue/Mouth
- Ext.Beak/Head/Neck
- Flippers/Shoulder/Armpit
- Carapace/Plastron
- Unknown,External
- Unknown
- Not Hooked

Exhibit O

Sea Turtle Demographics

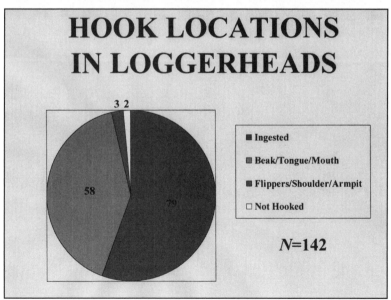

Exhibit O

Sea Turtle Demographics

251

HOOK REMOVAL IN LOGGERHEADS

Was the Hook Removed?		
	Yes	No
Ingested (*n*=79)	0	79
Beak/Tongue/Mouth (*n*=58)	38	20
External (*n*=3)	3	0

Exhibit O

Sea Turtle Demographics

UNITED STATES DEPARTMENT OF COMMERCE
National Oceanic and Atmospheric Administration
NATIONAL MARINE FISHERIES SERVICE
Silver Spring, MD 20910

NMFS PUBLISHES EMERGENCY RULE TO REDUCE SEA TURTLE BYCATCH AND BYCATCH MORTALITY IN HIGHLY MIGRATORY SPECIES FISHERIES

The National Marine Fisheries Service (NMFS) has issued an emergency rule to implement the reasonable and prudent alternative identified in the June 8, 2001, Biological Opinion to reduce bycatch and bycatch mortality of threatened loggerhead and endangered leatherback sea turtles. The closure and gear modifications required by the emergency rule affect all U.S. commercial fishermen who have been issued Federal highly migratory species (HMS) permits and use pelagic longline gear in the Atlantic Ocean, Gulf of Mexico, and Caribbean Sea. The emergency rule also requires ALL U.S. commercial and recreational fishermen that have been issued HMS fishing permits to post sea turtle handling and release guidelines in the wheelhouse. The rule is effective as of July 11, 2001 and will be in place through January 9, 2002. Comments will be accepted at the address listed below until

Area Closure

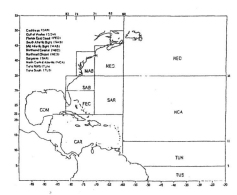

Source: Cramer and Adams, 2000.

The Northeast Distant Statistical Reporting Area (NED) will be closed effective **July 15, 2001, through January 9, 2002,** unless the rule is extended. The closed area is bounded by the following coordinates: 35°00' N. lat., 60°00' W. long.; 55°00' N. lat., 60°00' W. long.; 55°00' N. lat., 20°00' W. long.; 35°00' N. lat., 20°00' W. long.; 35°00' N. lat., 60°00' W. long. For the duration of this emergency rule, vessels issued Atlantic HMS permits are prohibited from fishing with pelagic longline gear in the NED.

Gear Modifications

Beginning August 1, 2001, through January 9, 2002, all vessels issued permits allowing pelagic longline fishing for HMS species will be required to modify how their gear is deployed. One

Exhibit P
NMFS Report of Sea Turtle Mortality

modification requires that gangions be attached to the mainline at least two gangion lengths from the floatline. The other modification requires that gangions be 10 percent longer than the floatlines, if the length of the floatline plus the length of the gangion is 100 meters or less.

Sea Turtle Handling and Release Guidelines

Effective September 15, 2001, through January 9, 2002, ALL Atlantic vessels that have been issued Federal HMS permits for any gear type and/or target species are required to post inside the wheelhouse NMFS issued guidelines for the safe handling of sea turtles. This measure will allow vessel captains to refer to the appropriate handling and release guidelines in the event a sea turtle is accidentally hooked or entangled. NMFS will distribute the guidelines via mail to all HMS permit holders and place the document on the Internet.

This FAX notice is a courtesy to the HMS fishery participants to help keep you informed about your fishery. Official notice of Federal fishing actions is made through filing such notice with the Office of Federal Register. For further information on this emergency rule and its requirements, contact Tyson Kade or Karyl Brewster-Geisz at 301-713-2347. Copies of the Environmental Assessment and Regulatory Impact Review and Emergency Rule are available per request from Highly Migratory Species Management Division, 1315 East-West Highway, Silver Spring, MD 20910 (phone: 301-713-2347, fax: 301-713-1917). The information will also be posted on the HMS website at: http://www.nmfs.noaa.gov/sfa/hmspg.html.

Bruce C. Morehead   7/11/01
Acting Director, Office of Sustainable Fisheries   Dated

Exhibit P

NMFS Report of Sea Turtle Mortality

include unidentified turtles not listed in the table. Most marine turtles were caught from the Grand Banks (NED) fishing area, outside of the U.S. EEZ. These estimates include the loggerhead, leatherback, Kemp's ridley. hawksbill and green sea turtles (see Appendix III). However, the records of the Kemp's ridley, hawksbill and green captures may have been misidentifications and should be re-evaluated (see Hoey 1998; Witzell 1999).

For 1998, Yeung (1999) provided estimates for the number of sea turtles "seriously injured", *i.e.*, those not expected to survive. Pooling across species but stratified by area, an estimated total of 730 sea turtles were taken (this estimate differs from the estimate provided in Table 2, below, because of differences in the procedures used to sum the estimates). Of these, Yeung (1999) estimates that all but 10 were seriously injured. This is a much greater predicted mortality rate than that reported by Aguilar *et al.* (1992). Yeung's (1999) criteria for determining serious injury were based on criteria developed for marine mammals (Angliss and DeMaster,1998) and may be overly conservative for sea turtles. These values still use the "old" methods of estimation (*i.e.* data were not pooled across quarters, years or areas).

Table 2. Estimated Sea Turtle Takes Recorded in the U.S. Atlantic and Gulf of Mexico Pelagic Longline Fishery for Swordfish, Tuna and Sharks, 1992 - 1998 (based on estimates in Johnson *et.al.*, 1999 and Yeung, 1999b, summed from estimates stratified by species and area).

Species	Loggerhead		Leatherback		Green		Hawksbill		Kemp's		Sum Total**
Year	Total	Dead*	Total	Dead*	Total	Dead*	Total	Dead*	Total	Dead*	
1992	247	18	871	87	129	18	30	0	0	0	1295
1993	374	9	889	12	25	0	0	0	0	0	1315
1994	1279	12	700	12	24	0	0	0	15	0	2047
1995	2169	0	925	0	31	0	0	0	0	0	3290
1996	410	0	674	0	0	0	0	0	0	0	1084
1997	329	0	357	0	0	0	13	0	23	0	765
1998	472	0	169	0	0	0	77	0	0	0	718

* Does not account for death that may occur after release, which several studies have shown to be 29-33%
**Totals include unidentified turtles not listed in the table.

Numbers for takes estimated using the preferred pooling order (quarter, year, area) and number of pooled samples (n = 5) selected by Yeung *et al.* (in prep.) are discussed hereafter, although summary data across years for all methods reported, where pooling of data was performed to extrapolate take over unobserved areas, yielded similar results. Total take reported for loggerheads, over the period 1992 - 1998, was 6,544, with a lower confidence interval of 3,152 and an upper confidence interval of 15,866 for the seven-year total. Totals for the most recent year available (1998) yield an estimate of 987 loggerheads taken (95% CI = 354 - 2,866). For leatherbacks, an estimated total of 5,003 turtles were taken over the same time-period, with lower and upper confidence limits of 2,014 and 14,420, respectively. For 1998, there were an estimated 394 leatherbacks taken (95% CI = 117 - 1,408). Estimates for 1999 are likely to be considerably higher than these estimates.

Exhibit P
NMFS Report of Sea Turtle Mortality

that time. Therefore, for 2000, the take levels for loggerheads and leatherbacks anticipated are based on 50% of the average annual estimated incidental level from logbook and observer data reported in Yeung and Epperly (in prep.), for loggerheads and leatherbacks. The annual incidental take levels for animals dead or sustaining life threatening injuries (*i.e.* hooked by ingestion or moribund upon release) are anticipated based on a projected 30% mortality rate and 5% observer coverage (if the required 5% level is not achieved, resulting numbers will be scaled accordingly): In 2001, when take prevention measures can be implemented over the entire year, takes must be further minimized to achieve a total reduction in take of 75% of the current average estimated levels. Take levels for Kemp's ridleys, green turtles and hawksbill turtles are not anticipated to change substantially from current take estimates, because one turtle extrapolated across total effort is the general case. Therefore, anticipated incidental take levels for sea turtles are as follows:

2000:
(a)   358 leatherback turtles entangled or hooked (annual estimated number) of which no more than six (6) are observed hooked by ingestion or moribund when released,
(b)   468 loggerhead turtles entangled or hooked (annual estimated number); of which no more than seven (7) may be observed hooked by ingestion or moribund when released,
(c)   46 green turtles entangled or hooked (annual estimated number) of which no more than two (2) can be observed hooked by ingestion or moribund upon release,
(d)   23 Kemp's ridley turtles entangled or hooked (annual estimated number), of which no more than one (1) can be observed hooked by ingestion or moribund when released, and
(e)   46 hawksbill turtles entangled or hooked (annual estimated number) of which no more than two (2) can be observed hooked by ingestion or moribund upon release.

2001:
(a)   179 leatherback turtles entangled or hooked (annual estimated number) of which no more than three (3) are observed hooked by ingestion or moribund when released,
(b)   234 loggerhead turtles entangled or hooked (annual estimated number); of which no more than four (4) may be hooked by ingestion or observed moribund when released,
(c)   46 green turtles entangled or hooked (annual estimated number) of which no more than two (2) can be observed hooked by ingestion or moribund upon release,
(d)   23 Kemp's ridley entangled or hooked (annual estimated number), of which no more than one (1) can be observed hooked by ingestion or moribund when released, and
(e)   46 hawksbill entangled or hooked (annual estimated number) of which no more than two (2) can be observed hooked by ingestion or moribund upon release.

The proposed action affects only the longline portion of the HMS fisheries, so the levels of incidental take that were anticipated in previous Opinions on HMS fisheries are not expected to change. Therefore, anticipated levels of incidental take for other HMS fisheries remain unaltered (as follows below). However, according to the most recent information available, a shark gillnet fishery may be forming off Alabama. Levels of incidental take for shark gillnet gear anticipated under this Opinion were formed without consideration of this additional effort. If additional effort takes place in this fishery under the purview of NMS (*i.e.* fishers hold a limited access permit for sharks), it must be monitored, and appropriate incidental take levels incorporated into a reinitiated opinion.

90

Exhibit P
NMFS Report of Sea Turtle Mortality

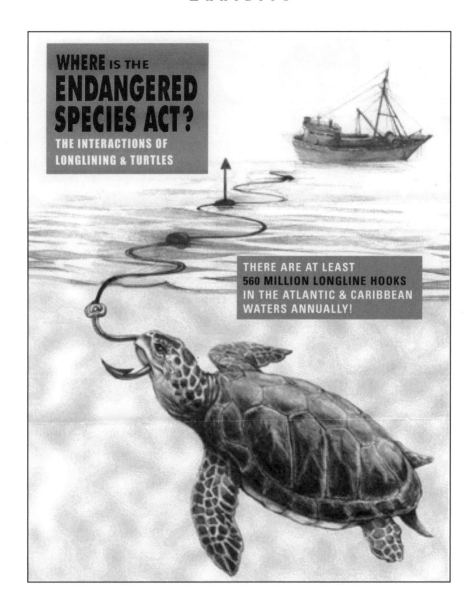

Exhibit P

Sloan's Turtle Campaign Poster

Schedule of ICCAT Species Yearly Catch

	M. Tons	Pounds	$ per lb.	$ Value
E Atlantic Bluefin Tuna	18,000	39,600,000	$5	$198,000,000
EA BFT (2nd cannery)	18,000	39,600,000	$2	79,200,000
W Atlantic Bluefin Tuna	2,600	5,720,000	$6	34,320,000
S Atlantic Bluefin Tuna	1,228	2,701,600	$3	8,104,800
N Atlantic Swordfish	10,600	23,320,000	$4	93,280,000
S Atlantic Swordfish	13,486	29,669,200	$4	118,676,800
N Atlantic ALBA	30,000	66,000,000	$1.5	99,000,000
S Atlantic ALBA	27,500	60,500,000	$1.5	90,750,000
SK Tuna	133,181	292,998,200	$.5	146,499,100
Blue Marlin	3,155	6,941,000	$1.5	10,411,500
White Marlin	1,022	2,248,400	$1.5	3,372,600
Bigeye Tuna	94,786	208,529,200	$4	834,116,800
Yellowfin Tuna	147,434	324,354,800	$2	648,709,600
Total Value	**500,992**			**$2,364,441,200**

Exhibit Q
Schedule of ICCAT Catch

MEMORANDUM

To: American Delegation at ICCAT Intersessional

From: Stephen Sloan

Date: April 2001

Re: Compliance:
       Creation of ICCAT Fishing Flag

BACKGROUND:

ICCAT has been in existence since 1969: contracting parties (CCP) and cooperating parties (COOP) are required to report their species landings annually to SCRS. Any violation by a CCP/COOP may subject the violator to being declared an I.U.U. party. If so sanctioned, the violator will not be able to sell the catch to another CCP or COOP. This is both an internal and external economic sanction approved by the CCP and COOP members of ICCAT.

Notwithstanding the contract among the CCP and COOP ICCAT members, it has been increasingly difficult to obtain landing statistics because of Flags of Convenience (FOC) methods applied to the registry of fishing vessels (F/V).

For example: A boat built in Spain, owned by a Spanish corporation, may be reflagged in Panama for a variety of economic and insurance reasons. That vessel may unload its catch in a non-Spanish or non-Panamanian port, say Abidjan in Ivory Coast. The catch of that vessel and the catch report due SCRS raises the question: To whom is the catch allocated? (a) Spain; (b) Panama; (c) Ivory Coast; (d) to no one?

If no SCRS landings report is issued, clearly this vessel, whose FOC registration-ownership is now in question because of the reflagging, becomes an I.U.U. fishing vessel. Yet, even if the landing report is issued, to whose allocation does the catch belong? Clearly, FOC vessels complicate the compliance process at ICCAT.

Exhibit R

Sloan Memorandum to ICCAT of April 2001

Index

270